Electrodynamic Characteristics of Accelerating Cavities

Electrodynamic Characteristics of Accelerating Cavities

N.P. Sobenin and B.V. Zverev
Moscow Engineering Physics Institute, Russia

Translated from the Russian by Zyufyar Feizulin

**MOSCOW ENGINEERING PHYSICS INSTITUTE
UNIVERSITY PRESS**
FOUNDATION FOR INTERNATIONAL SCIENTIFIC AND EDUCATION COOPERATION

London • Moscow

Originally published in Russian in 1993 as ЭЛЕКТРОДИНАМИЧЕСКИЕ ХАРАКТЕРИСТИКИ УСКОРЯЮЩИХ РЕЗОНАТОРОВ by Energoatomizdat, Moscow
© 1993, The authors

Amsteldijk 166
1st Floor
1079 LH Amsterdam
The Netherlands

British Library Cataloguing in Publication Data

Sobenin, N. P.
    Electrodynamic characteristics of accelerating cavities
    1.Cavity resonators 2.Linear accelerators 3.Ion flow dynamics
    I.Title II.Zverev, B. V.
    539.7′33

ISBN: 90-6994-003-5

Contents

PREFACE

Resonance accelerators of charged particles are increasingly widely used in industry, medicine, and physics. Thousands of standing wave linear accelerators (linacs) built around biperiodic accelerating structures are being used all over the world. Accelerating structures are used in linear ion accelerators, in storage rings, in race track microtrons, in storage-stretcher rings, and in other similar systems. The efficiency of these systems, their cost, dimensions, and reliability are defined largely by the electrodynamic characteristics and design of their accelerating cavities.

Some books on accelerating systems cover the electrodynamic design of the disk loaded waveguide, which is the central element of traveling wave linacs [1–4]. Various aspects of accelerating cavity design are considered in numerous research papers, and a recent upsurge of interest in the methods of electrodynamic modeling of resonators has increased the number of papers devoted primarily to mathematical or general theoretical issues of modeling resonators and accelerating structures [5, 6]. Separate issues related to the dynamic range of the cavity, specifically the particle dynamics in cavity accelerators, may be found in miscellaneous publications [7–10].

Unfortunately, the measurement aspects of investigations into the electrodynamic characteristics of cells and tuning of accelerating multicell cavities have been poorly reported [3, 11].

This book systematizes the material concerning numerical and analytical design, experimental investigations, and tuning of *E*- and *H*-wave driven accelerating structures. It is mainly based on the results obtained in the development of small electron and proton linacs. For example, E-type accelerating structures are designed using electrodynamic modeling methods, and the calculation of electrodynamic characteristics of H-type cavities is based on the analog resonator method, which yields analytical expressions.

The potential of the suggested methods is demonstrated by the electrodynamic design of both known biperiodic accelerating structures and interdigital H-type resonators, and new types of structures. The results of these calculations are confirmed by experiments with mock-up cavities equipped with probes.

Cell coupling and cavity–rf generator coupling are important issues in the design of multicavity systems. Since such systems are axially asymmetric, they are hard to deal with using numerical methods. The absence of relevant design techniques in the literature prompted us to develop approximate methods for designing coupling elements in biperiodic structures and the waveguiding and coaxial junctions with TM- and TE-wave driven cavities.

This book summarizes relevant design material in a form suitable for practical applications in the development of accelerating cavities. This is illustrated by the design of small standing wave linacs using biperiodic structures (RELUS series accelerators) and a compact proton H-type linac built around a cylindrical interdigital cavity (Uragan-2 accelerator).

Since the Russian edition of this book was published almost a decade ago, some quoted reviews are no longer exhaustive, but we have kept them in this edition as historical hallmarks.

We have included in Chapter 4 three new sections reflecting our recent studies. One is connected with the development of an accelerating structure (coupler and dipole mode investigation) for the linear electron-positron collider of the DESY (Deutsches Elektron Synchrotron, Hamburg, Germany). The other two cover the development of a prismatic biperiodic structure with good accelerating and focusing characteristics for a portable race track microtron. This material replaces sections devoted to the electrical strength of accelerating structures, an area which is covered rigorously in the existing literature.

We express our sincere gratitude to all our colleagues in the Moscow Engineering Physics Institute who assisted us in writing this book.

N.P. Sobenin and B.V. Zverev

1

ACCELERATING STRUCTURES OF LINEAR ACCELERATORS

1.1 Electrodynamic Characteristics

In choosing an accelerating structure producing a required beam
of charged particles, one has to solve equations describing the
electromagnetic fields and the motion of charged particles in
these fields. This problem is usually split into equations of
electrodynamics and equations of particle motion. Having solved
Maxwell's equations or having used available measurement data
one determines the electrodynamic characteristics (EDC) which
are then used in the calculation of particle dynamics [1, 2, 8, 9,
12].

Two types of operation charged particle accelerators are
common, namely, the traveling (TW) and standing wave (SW)
operations. In the TW case, the electrodynamic structure is an
accelerating waveguide; in the SW case, an accelerating resonator
in the form of a single cavity or a chain of coupled cavities.

The main electrodynamic characteristics of accelerating waveguides include the dispersion characteristic, attenuation coefficient, shunt impedance, and overvoltage factor.

The dispersion characteristic relates the wave phase velocity v_{ph} to the rf frequency f for the given dimensions of the accelerating waveguide. The dispersion dependence can be determined theoretically or found experimentally [3].

It is convenient to plot the dispersion characteristic against $1/\lambda = F(1/\lambda_{\text{wg}})$ or $f = F(\theta)$, where λ and λ_{wg} are the wavelengths in the free space and waveguide, respectively; θ is the mode or phase advance over a period D of the structure with

$$\theta = k_z D = 2\pi D / \lambda \beta_{\text{ph}}. \tag{1.1}$$

Entering the plot represented as a function of $1/\lambda = F(1/\lambda_{\text{wg}})$, one can readily determine the relative phase velocity

$$\beta_{\text{ph},m} = \frac{v_{\text{ph},m}}{c} = \frac{2\pi/\lambda}{2\pi/\lambda_{\text{wg},m}} = \tan \varphi_m, \tag{1.2}$$

the relative group velocity

$$\beta_{g,m} = \frac{v_{g,m}}{c} = \frac{1}{c}\frac{d\omega}{dk_{zm}} = \frac{d(1/\lambda)}{d(1/\lambda_{\text{wg}})} = \tan \psi_m, \tag{1.3}$$

the dispersion parameter

$$k_{\text{d},m} = \frac{\lambda}{v_{\text{ph},m}}\frac{dv_{\text{ph},m}}{d\lambda} = \frac{\beta_{\text{ph},m}}{\beta_{g,m}} - 1, \tag{1.4}$$

and the coupling coefficient

$$k_c = \frac{\left|f_\pi^2 - f_0^2\right|}{f_\pi^2 + f_0^2}$$

$$\approx 2\frac{\left|f_\pi - f_0\right|}{f_\pi + f_0} \text{ for small } k_c. \tag{1.5}$$

Here, φ_m is the angle made with the abscissa by a straight line from the origin to a point on the dispersion curve of a selected frequency, and ψ_m is the angle made with the abscissa by the tangent to the dispersion curve at a point corresponding to the selected frequency; f_π and f_0 are the frequencies of the π and 0 modes, respectively; and index m here and above indicates the number of the spatial harmonic emerging by representing any periodic field component in the form

$$E_z = \sum_{m=-\infty}^{\infty} E_{zm} e^{-\alpha z} e^{-ik_{zm}z}, \qquad (1.6)$$

where E_{zm} is the longitudinal electric field of harmonic m, and $k_{zm} = (\theta + 2\pi m)/D$ is the longitudinal wave number of the mth spatial harmonic.

The attenuation of the electric field may be expressed in terms of the power flux in the waveguide P and the loss of power per unit length $P_1 = |\,dP/dz\,|$ as

$$\alpha = \frac{1}{2}\frac{P_1}{P}, \qquad (1.7)$$

where

$$\left|\frac{dP}{dz}\right| = \frac{R_w}{2D}\int_S H^2 ds, \qquad (1.8)$$

$$R_w = \pi Z_0 \delta/\lambda, \qquad (1.9)$$

$$\delta = \sqrt{\frac{2}{\omega \sigma_w \mu_0 \mu}}, \qquad (1.10)$$

$Z_0 = \sqrt{\mu_0/\varepsilon_0}$ is the wave impedance of free space, R_w is the wall surface resistance, δ is the penetration depth, σ_w is the conductivity of the accelerating waveguide wall, and S is the cell area of the accelerating waveguide with period D.

The shunt impedance per unit length characterizes the efficiency of the structure in providing a high accelerating field for a given rf power dissipation in the structure. It is defined as

$$r_{sh} = \frac{1}{PL}\left[\int |E(z)|dz\right]^2,$$
(1.11)

where $E(z)$ is the complex amplitude of the electric field on the axis and P is the total power dissipated in the structure. Since $P_1 = |dP/dt|$, this relation may be rewritten as

$$r_{sh,\,m} = \frac{E_{zm}^2}{|dP/dz|} = \frac{E_{zm}^2}{2\alpha P},$$
(1.12)

where E_{zm} is the amplitude of the mth harmonic of the longitudinal component of the electric field on the axis of the structure.

It should be noticed that sometimes the shunt impedance for accelerators is defined as in circuit theory, which adds another factor 2 in the denominator of (1.12).

The coupling impedance of the accelerating structure

$$R_c = \frac{E_z^2}{2k_z^2 P} = \frac{1}{2k_z^2}\frac{\omega}{v_g}\frac{r_{sh}}{Q},$$
(1.13)

where Q is the quality factor related to the attenuation (on the assumption that the group velocity and energy propagation velocity are identical) by

$$Q = \frac{\omega}{2v_g\alpha}.$$
(1.14)

A transverse stability analysis of accelerators calls for the transverse shunt impedance per unit length

$$r_{sh\perp} = \frac{1}{2\alpha P}\left(\frac{1}{k}\frac{\partial E_z}{\partial x}\right)^2,$$
(1.15)

where x is the transverse coordinate in the HE_{11} wave polarization plane, E_z is the amplitude of the synchronous harmonic of the longitudinal electric field in the HE_{11} wave polarization plane.

The electric field overvoltage factor is

$$k_{ov} = \frac{E_{s,\max}}{E_{z0}}, \qquad (1.16)$$

where $E_{s,\max}$ is the maximum strength of the electric field on the surface of the accelerating waveguide, and E_{z0} is the amplitude of the fundamental spatial harmonic of the longitudinal electric field on the waveguide axis.

The time for electromagnetic energy to fill a accelerating section of length L is given by the formula

$$t_f = \int_0^L \frac{dz}{v_g(z)}. \qquad (1.17)$$

Clearly, for a section with invariable group velocity, $t_f = L/v_g$.

Among other characteristics of accelerating waveguides, we note the impedance characteristic which reflects the dependence of the input impedance Z_{in} on frequency. The input impedance is related to the reflection coefficient at the waveguide input, Γ_{in}, by the known expression

$$\frac{Z_{in}}{Z_0} = \frac{1 + \Gamma_{in}}{1 - \Gamma_{in}}. \qquad (1.18)$$

The main electrodynamic characteristics of accelerating cavities are the resonance frequency, dispersion characteristic (in the case of a chain of coupled cavities), unloaded quality factor, cavity-generator coupling factor, and effective shunt impedance.

In the absence of losses in the walls and a cavity-filling medium, the resonant frequency of any mode is a real valued quantity. In resonators of intricate geometry, these frequencies are determined by numerical methods or experimentally.

The unloaded quality factor of the accelerating cavity is

$$Q_0 = \frac{\omega W}{P} = \frac{\omega W_1}{|dP/dz|},$$ (1.19)

where W_1 is the energy per unit length, and W is the maximum amount of em energy in the cavity of volume V,

$$W = \frac{\varepsilon_0}{2} \int_V |\mathbf{E}|^2 dv.$$ (1.20)

The cavity-generator coupling factor β_0 is the ratio of the unloaded quality factor to the external quality Q_{ext}:

$$\beta_0 = \frac{Q_0}{Q_{ext}}.$$ (1.21)

As a rule, the coupling factor of the beam-loaded cavity must be close to unity (critical coupling) because, in this case, the power transmitted from the generator to the resonator is a maximum. The power transmission factor is

$$k_t = 1 - |\Gamma_{in}|^2 = \frac{k_{t0}}{1 + a_1^2},$$ (1.22)

where Γ_{in} is the reflection coefficient at the cavity input,

$$k_{t0} = \frac{4\beta_0}{(1 + \beta_0)^2}$$ (1.23)

is the transmission coefficient at the resonance frequency ω_0,

$$a_1 = 2Q_1 \frac{\Delta\omega}{\omega_0}$$ (1.24)

is the generalized detuning of the loaded resonator, $\Delta\omega = \omega - \omega_0$ is the frequency detuning, and Q_1 is the loaded quality factor given by the formula

$$\frac{1}{Q_1} = \frac{1}{Q_0} + \frac{1}{Q_{\text{ext}}}. \tag{1.25}$$

The effective shunt impedance in the accelerating cavity $R_{\text{sh,eff}}$ relates the equivalent voltage U between two points in the cavity over a given path to power P dissipated in the cavity walls

$$R_{\text{sh, eff}} = \frac{U^2}{P},$$

where

$$U = |\dot{U}| = \left| \int_{z_1}^{z_2} E_z(z) \exp(ik_z z)\,dz \right|, \tag{1.26}$$

$E_z(z)$ is the distribution of the longitudinal component of the electric field over the cavity axis, and z_1 and z_2 are points on cavity axis such that $z_2 - z_1 = L$ is the cavity length.

The effective shunt impedance (1.26) takes into account the field variation in the cavity during the passage of accelerated particles. The transit-time factor is given by

$$T = \frac{\left| \int_0^L E_z(z) \exp(ik_z z)\,dz \right|}{\int_0^L |E_z(z)|\,dz} \tag{1.27}$$

with

$$R_{\text{sh, ef}} = R_{\text{sh}} T^2 \tag{1.28}$$

where the shunt impedance of the accelerating cavity is

$$R_{\text{sh}} = \frac{U_0^2}{P} = \frac{1}{P} \left(\int_0^L |E_z(z)|\,dz \right)^2. \tag{1.29}$$

The effective shunt impedance per unit length $r_{sh,eff}$ is given as follows:

$$r_{sh,eff} = \frac{1}{PL}\left|\int_0^L E_z(z)\exp(ik_z z)dz\right|^2.$$

(1.30)

Using the expressions for shunt impedance and quality factor we may write the characteristic impedance as

$$\frac{R_{sh}}{Q} = \frac{1}{\omega W}\left(\int_0^L |E_z(z)|dz\right)^2$$

(1.31)

or

$$\frac{R_{sh,eff}}{Q} = \frac{1}{\omega W}\left|\int_0^L E_z(z)\exp(ik_z z)dz\right|^2.$$

(1.32)

The ratio R_{sh}/Q depends on the cavity geometry and does not depend on the cavity's material and quality of element matching. This ratio is sometimes called the characteristic impedance.

In the case of accelerating resonators, the electric field overvoltage factor may be introduced as the ratio of $E_{s,max}$ to the product of the mean longitudinal electric field $\overline{E}_z$ by the transit-time factor, viz.,

$$k_{ov} = \frac{E_{s,max}}{\overline{E}_z T}.$$

(1.33)

One more important quantity is the filling time of the cavity with rf power. The filling time constant is given by

$$t_f = 2Q_1/\omega.$$

(1.34)

The field attains 80% of its steady state value in $1.6t_F$ and 99% in $4.6t_F$.

In design of high intensity accelerators, it is important to estimate the effect of higher order modes on the accelerating process. In this case, it is convenient to use the loss factor [100]

$$k_{l_n} = \sum_n \frac{\omega_n}{4} \left(\frac{R_{sh,eff}}{Q} \right)_n F_n,$$ (1.35)

where F_n is the form factor depending on the distribution of the linear charge density. For a Gaussian distribution,

$$F_n = \exp\left(-\frac{\omega_n^2 \sigma_z^2}{c^2} \right),$$ (1.36)

where σ_z is the halved rms length of the bunch ($\sigma_z/c = \sigma_t$).

For engineering analyses and design of accelerating systems, it is important to know how the electrodynamic characteristics vary with frequency. We used such dependencies to tabulate the characteristics of disk loaded waveguides (DLW) and biperiodic structures (BPS) in our handbook [3]. Accordingly, the shunt impedance per unit length varies as $\omega^{1/2}$, the quality factor as $\omega^{-1/2}$, the ratio r_{sh}/Q as ω, and the attenuation factor α as $\omega^{3/2}$.

1.2   Electron Accelerating Structures

For linear electron accelerators, a widely accepted accelerating system is the disk-loaded waveguide (DLW). Standing wave accelerators use biperiodic structures (BPS).

Figure 1.1 shows the cross section of a DLW. This design is attractive because of its high shunt impedance and amenability to mass production.

The function of specific shunt impedance of the mode in a DLW exhibits a shallow maximum near $\pi/2$ and $2\pi/3$ [3]. In order to reduce the effect of beam break-up and the number of disks in the waveguide, the $2\pi/3$ mode is preferred. DLWs excited in $3\pi/4$

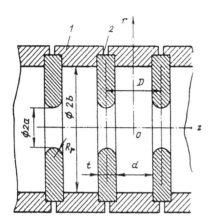

Figure 1.1. Sectional view of an disk loaded waveguide: (1) ring, (2) disk

and $4\pi/5$ modes have the same advantages; however, these configurations are hard to adjust and have a slow group velocity, which imparts the implementation of linacs with these modes.

Rounding the internal surface of the cell, as shown in Fig. 1.2a, reduces the rf losses in the walls, thus increasing the shunt impedance of the DLW. As a result, in the $2\pi/3$ mode, r_{sh} increases by 10–12% compared to r_{sh} values characteristic of traditional geometries. A further increase of $r_{sh,eff}$ is achieved by using a nose-cone near the aperture (Fig. 1.2b). This approach increases the shunt impedance for the $7\pi/8$ mode by 30% compared to the initial cell geometry. However, this geometry noticeably decreases the group velocity [14]. To increase the group velocity while preserving $r_{sh,eff}$, one should provide additional elements that improve the magnetic field coupling. These elements may be configured as narrow slits in the disks, as shown in Fig. 1.3. Table 1.1 summarizes the specific effective shunt impedance as a function of the group velocity for DLWs with magnetically coupled cells for the $4\pi/5$ mode, and with electrically coupled cells for the $2\pi/3$ mode. Clearly, for backward wave DLWs, where the region of rf cell coupling is separated from the region where particles interact with the field, for a given v_g, one can obtain a considerably higher $r_{sh,eff}$ than for forward wave DLWs [15].

Figure 1.2. Sectional view of the top part of two DLW cells.

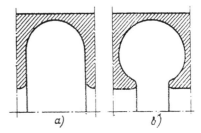

Figure 1.3. Disk loaded tube with magnetic slots.

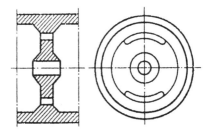

Table 1.1 Effective relative shunt impedance per unit length $r_{\text{sh,eff}}$ as a function of β_g for disk loaded waveguides with electrical and magnetic coupling [15]

Electric coupling ($\theta = 2\pi/3$)		Magnetic coupling ($\theta = 4\pi/5$)	
β_g	$r_{\text{sh,eff}}$, MΩ/m	β_g	$r_{\text{sh,eff}}$, MΩ/m
0.006	72.3	0.007	95.6
0.009	69.1	0.011	92.0
0.014	66.0	0.015	88.6
0.019	62.8	0.020	85.0
0.021	61.2	0.028	81.4
0.024	55.5	0.045	76.4
		0.056	73.4
		0.061	69.9
		0.071	65.7

We note that the choice of a specific mode for a backward wave structure between 2π and π is a tradeoff between a high $r_{\text{sh,eff}}$ and a sufficient coupling for different dimensions of coupling slots. One should also take into account the difficulties of waveguide tuneup. For example, for a backward traveling-wave linac, Trong [16] choose the $4\pi/5$ mode.

For standing wave linacs, the DLW is no longer an effective accelerating structure. In SW structures, the shunt impedance is equal to the product of the shunt impedance of a cell to the number of cells at 0 and π phase shifts only. At all other modes, this shunt impedance is half as large [12].

The drawbacks of SW linacs operating with 0 or π modes include a zero group velocity, poor frequency separation of these modes from the adjacent modes, and, as a consequence, an unstable operation of the linac.

Clearly, it is desirable to obtain an accelerating structure which would have a high shunt impedance and a high stability typical of the $\pi/2$ mode. These features are combined in the biperiodic structure (BPS). With $\pi/2$ mode fields, every other cavity cell is unexcited. Obviously, a considerable reduction of length of such unexcited (coupling) cells does not change the field distribution for the $\pi/2$ mode. However, with respect to the accelerated beam, this structure may be viewed as exciting oscillations close to the π mode, because shorter coupling cells or the displacement of these cells to the periphery considerably increases the shunt impedance.

Cells are coupled via the magnetic field. A further increase of the shunt impedance is achieved by optimizing the accelerating cell geometry into an Ω shape, as illustrated in Fig. 1.2b. A characteristic feature of this configuration is the presence of drift tubes with rounded ends which increase the intensity of the electric field in the beam channel. This increase is achieved by changing the gap between the drift tubes. A smaller gap increases the transit time factor T, but decreases r_{sh}/Q. Therefore, a certain gap length yields the optimum value $r_{\text{sh,eff}}/Q = (r_{\text{sh}}/Q)T^2$. The drift tube geometry affects r_{sh}/Q in a lesser degree; therefore, its choice is governed mainly by the electric strength of the accelerating cell because the field on the surface is a maximum in the nose segment of the drift tube.

Several designs of the biperiodic accelerating structure are presented in Fig. 1.4. The performance of these structures may be compared in terms of their electrodynamic characteristics (EDC), primarily in terms of $r_{sh,eff}$, k_c, and the presence of parasitic oscillations near the operating frequency. On the other hand, they may be compared in their weight and size performance figures, amenability to mass production, vacuum conductivity, and heat removal capacity. Designs (a), (c) and (e) have an enlarged effective diameter, whereas those in (b) and (d) have enlarged gaps between accelerating cells and, thus, a reduced $r_{sh,eff}$. However, optimizing the dimensions of a biperiodic structure one can achieve a comparatively high values of $r_{sh,eff}$ for all these designs. Structures (b) and (d) with their cylindrical symmetry offer certain technological advantages because they do without preliminary

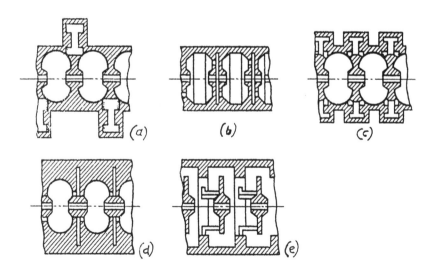

Figure 1.4. Biperiodic accelerating structures: (a) with side coupling cells, (b) with on-axis coupling cells, (c) with annular ring coupling cells, (d) with coaxial coupling cells, and (e) disc-and-washer structure.

brazing of numerous small components. Designs with coaxial coupling cells are preferable to other biperiodic structures because of a reduced risk of exciting higher order modes (HOMs) responsible for a transverse instability of the beam. This is an obvious advantage over BPSs with internal coupling cells, because in structures with coaxial coupling cells, the beam does not interact with these cells. As to the structures with external coupling cells, experiments indicate that at higher coupling coefficients, the frequency of the dipole mode for this type of structure can be comparable with the operating frequency [17].

Andreev [18] set forth one more configuration of the biperiodic structure, namely, a disk-and-washer structure (DAW) depicted in Fig 1.4e. The space between two adjacent washers may be viewed as an accelerating cell, while that between two disks as a coupling cell. The DAW has a characteristically high coupling coefficient (up to 0.5), a high effective shunt impedance, and a high vacuum conductivity. On the other hand, its ratio $r_{sh,eff}/Q$ is somewhat lower than for the considered biperiodic structures, since, at the same field strength on the axis, the energy stored in the cavity proves to be somewhat higher owing to a considerable transverse size of the DAW. However, the quality factor is high since the displacement current between washers and disks replace some conductivity currents on the external surface. As a result, $r_{sh,eff}$ proves to be high.

The drawbacks of DAWs are associated with the fact that the fundamental mode is not the lowest mode. Its frequency range includes HOMs. The pattern is aggravated by the fact that the stems favor additional parasitic modes. Reportedly, these HOMs can be removed by non-resonance or resonance methods [19, 154].

In order to compare the dynamic ranges of DLW and some BPSs, Table 1.2 summarizes the respective data at a frequency of 2856 MHz [19]. Its first and second columns list the EDCs for biperiodic structures with internal and external coupling cells, respectively. Clearly, if one reduces the beam aperture radius in the DAW to the values of biperiodic structures, their values of $r_{sh,eff}$ become equal.

Table 1.2 EDCs of accelerating structures

Accelerating structure	BPSs		DAW	DLW
Mode	$\pi/2$	$\pi/2$	π	$2\pi/3$
Q_0	15300	16208	20000	13200
$r_{sh,eff}$, MΩ/m	77	79	65	38.4
β_{gr}	0.05	0.026	0.5	0.012
Aperture radius, cm	054	0.6	0.75	1.13
Cavity radius, cm	3.94		6.3	4.13

The considered disk-loaded and biperiodic structures are widely used in various electron accelerators. A detailed analysis of the status of linacs based on DLWs and BPSs may be found in our handbook [3]. At present, several thousand linacs are used in different application fields such as ray therapy of malignant tumors, nondestructive testing of thick metal products, activation analysis, and radiation technology.

It is obvious that the requirements posed on the accelerating structures of linacs designed for fundamental research differ from those imposed on application systems. For example, it is required that the accelerating structures of electron-positron colliders for 2×0.5 or 2×1 TeV [20–23] should operate at extremely high accelerating fields (up to 100 MV/m).

Most known linear colliders use DLWs as accelerating structures [24–25]. Its simple technology and satisfactory dynamic range make the DLW a more promising design for this application.

The drawbacks of the DLW are associated primarily with a relatively small group velocity and, as a consequence, with more strict dimensional tolerances, and the need for a higher average rf power.

One more disadvantage of DLWs used in linear colliders is a relatively small iris aperture which leads to transverse instability due to wake fields. This drawback is somewhat weaker in structures based on smooth waveguides with periodic longitudinal excitation, such as zigzag bellows-type structures and phase-advance structures [27].

The most popular application area of biperiodic structures is compact SW electron accelerators for ray therapy and nondestructive testing. However, biperiodic structures are at advantage in systems with circulating continuous beams, as in race track microtrons. These applications require a high stability of the amplitude and phase of the accelerating field, a possibility to accelerate high intensities, and a developed surface for heat dissipation. These requirements are met by biperiodic structures with a large coupling factor, good separation of the main mode and HOMs, and a traditionally high shunt impedance. It is common practice also to choose a moderate gradient of the accelerating field (up to 1.5 MV/m) because of a need for high rf powers and stabilization problems at high gradients.

Race track microtrons with biperiodic accelerating structures were built in the Institute of Nuclear Physics at the Moscow State University [28] and in Mainz, Germany [29]. The first of these systems uses a 2450 MHz biperiodic structure with external coupling cells.

Continuous high-energy electron beams can be obtained in pulse TW or SW linacs with a storage stretcher ring. Particles are slowly extracted from this ring in the time interval between two sequential injection pulses. The synchrotron radiation and dissipation losses of the electron beam in the storage ring are compensated by accelerating cavities. Since such cavities operate with a high current load, the designer has to minimize the beam losses at HOMs. In this connection, it is important to reduce the number of nonoperable modes in the frequency range below the cutoff of the vacuum chamber or to damp these modes. At high current loads the shunt impedance is no longer a critical consideration.

A preferable accelerating system operating in combination with a stretcher storage ring is a chain of optimized slot coupled cavities operating in the π mode. These cavities are of the type used in cyclotrons and storage rings [30–32].

Current developments include accelerating systems with stretcher storage ring based on existing linacs or new structures. For example, the Kharkov Physicotechnical Institute, Ukraine, designed a stretcher storage ring based on a modified LUE-2000 linac [33]. Its accelerating cavities are designed as π mode slot coupled resonators for 699.3 MHz. The maximum circulating current is 160 mA in the pulse stretching mode, and up to 500 mA in the storage operating mode.

Acceleration of high average currents is one of the most important problems in linacs employed for the development of a laser on free electrons with a high average beam power and brightness. This problem is in turn associated with the need for accelerating cavities with the minimum beam power dissipated for the excitation of HOMs. The preferable solution for this task is a chain of uncoupled cavities [34] which is less sensitive to excitation of higher order modes due to the absence of direct coupling. These cavities may be configured as optimized designs with drift tubes (Fig. 1.2b) or with an enlarged beam channel without drift tubes (Fig. 1.2a). The last geometry requires a more powerful rf source; however, it is less affected by HOMs and these modes can be damped easier [35].

Linacs with hundreds of amperes in the bunch use the em field storage in the accelerating structure. Once the structure has been filled, it passes one or a few bunches of electrons which consume a considerable proportion of the stored energy. Since the stored energy stays in the structure for a short time and a considerable part is carried away by the electrons, one may expect a higher rate of acceleration from such accelerators due to small losses in the walls. Accelerating resonators with large amount of stored energy are used as sources of electrons and positrons in colliding-beam machines [36] and in electron storage rings [37].

They can be analyzed for the rf power requirement and electrical strength. None of the listed systems—DLW, biperiodic side coupled structures, and DAWs—has a considerable advantage over the others in terms of the rf overvoltage and energy filling efficiency [38]. However, the DAW has a certain advantage in terms of the coupling coefficient.

1.3   Accelerating Cavities for Ion Accelerators

At small phase velocities $(0.005 \leq \beta_{ph} \leq 0.1)$ the accelerating structures for ion linacs are built on the basis of lines with drift tubes and H-type resonators [7]. In the range of medium phase velocities $(0.04 \leq \beta_{ph} \leq 0.5)$, these are mainly Alvarez accelerators, and, for large β_{ph}, biperiodic accelerating structures are in wide use.

The choice of an accelerating structure is conditioned by such factors as the effective shunt impedance, accelerating rate, stability of the beam to variations in the geometry and regime of the rf generator, electrical strength, size, cost, amenability to mass production, and convenience in maintenance.

One of the oldest long line accelerating systems is the Sloan and Lawrence structure configured in the form of a twin line in a screen whose conductors are alternatively connected to drift tubes. This class includes also the Wideröe structure [39] with opposed drift tubes and some modifications of helical structures.

Lines with drift tubes multiply apply the potential difference between these lines. Figure 1.5 shows the Wideröe structure where the outer cylinder of the coaxial line is used as one of the

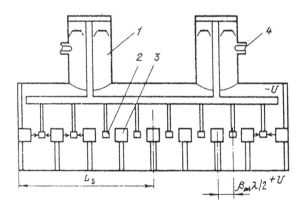

Figure 1.5. Wideröe structure: (1) stubs, (2) short drift tubes, (3) drift tubes with quadrupoles, and (4) coupling loops.

lines, and the internal conductor is somewhat offset from the center line. The line is loaded by stems carrying drift tubes. The electrical length of drift tubes is selected equal to the π or even 3π mode at the initial sections ($\theta = 2\pi D/\lambda\beta_{ph} = \pi$, i.e., $D = \lambda\beta_{ph}/2$). A length L_s of the structure with a stub in the middle is viewed as one section. Stubs are used to tune the system to resonance at a given frequency and to obtain a desired voltage pattern along the cavity axis. Loops in stubs are used to monitor the amplitude and phase of the field. The large diameter drift tubes house quadrupole elements. With four prestripping sections used in the UNILAC accelerator operating at 27.1 MHz, the effective shunt impedance varies from 33 to 65 MΩ/m and the rate of acceleration is 1 MeV/m [40].

Accelerating cavities with opposed vibrators provide for a more uniform longitudinal electric field than in the Wideröe structure. For example, in a cavity with two opposed vibrators (Fig. 1.6) the voltage antinode at one of them means a voltage node at the other, therefore, the voltage differential along the

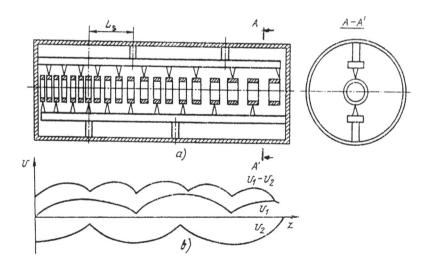

Figure 1.6. (a) Structure with opposed vibrators. (b) Corresponding electric field distribution along the cavity.

cavity is large and the voltage pattern is uniform.

Every drift tube is connected to the body by supports. The spacing between two adjacent supports of a vibrator is about $\lambda/2$ (with allowance for the electric "shortening" due to the effect of the capacity between the tubes and the capacities of the vibrators to the body) and the supports of one vibrators are shifted relative to the supports of the other by $\lambda/4$. The segment between two adjacent supports is taken as a section of the accelerating cavity, L_s.

In the initial section, where the drift tube spacing $L_s = \beta_{ph}\lambda/2$ is variable, a constant strength of the electric field is maintained by increasing the voltage along the cavity length. Thus, to increase the voltage at the cavity output one should shift the support to the cavity input, thus increasing the section length where the voltage needs to be increased.

Known spiral accelerating systems include horseshoe half-wave vibrator structures, one- and two flat-helix structures, and three-gap structures of the split-ring type. The main advantage of helix accelerating systems is the possibility of obtaining slow phase velocities at low frequencies in structures of moderate diameter. Superconducting helix systems can accelerate ions up to energies of 15 MeV/nucleon.

In a cylindrical cavity excited by a H-wave, the cavity is loaded by drift tubes which are alternatively connected to the opposite sides of the wall thus producing the desired accelerating field in the gaps. Figure 1.7 depicts an H-type cavity with individually suspended drift tubes [41]. Figure 1.8 shows a

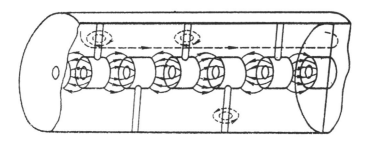

Figure 1.7. Interdigital structure of an H-type resonator.

modification of this structure where drift tubes are attached to two cam-like girders placed oppositely on the cavity wall short of its end surface [42]. The solid lines represent the electric field lines, and the dashed lines represent the magnetic field lines.

An important feature of H-type structures is a substantially reduced body diameter. While in the Alvarez structure excited by a TM_{010} mode the cavity diameter is $\lambda/1.3$, in the cavity with TE_{111} mode it can be reduced by a factor of 2 to 5. This reduction implies that such structures are especially well suited for heavy ions at relatively low rf frequencies. One more advantage of such structures is a high rate of acceleration due to the opposite field vectors in adjacent gaps (π mode). In order to take full advantage of this feature, one should provide a uniform distribution of the accelerating field along the structure.

One of the first linear accelerators based on the H-type cavity

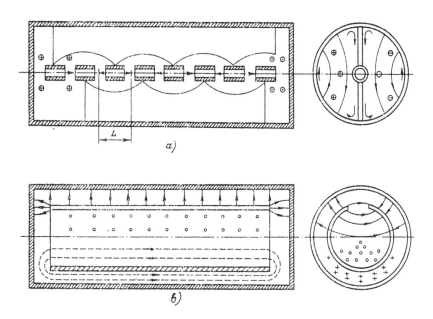

Figure 1.8. Two-chamber H-type resonators excited in (a) TE_{111} mode and (b) TE_{011} mode.

was the proton linac built in the Kharkov Physicotechnical Institute. It was a 2.5 MeV structure 0.92 m long and 0.3 m in diameter [43].

For the 8.5 MeV LUMZI accelerator, this team developed an accelerating structure resonant in a TE_{111} mode. They managed to achieve a uniform longitudinal field pattern without sacrificing other EDCs. This uniformity was achieved by changing the angle of "countercurrency" in a modification with symmetric adjustment stems ("ant" type structure) and a resonance element formed at the end segments of the structure (Fig. 1.9) [44].

A similar tuning element was used in a structure for ion accelerator with an injection energy of 0.24 MeV/nucleon and an output energy of 2.5 MeV built at the Tokyo Institute of Technology [45]. In this machine, the beam axis is offset from the shell axis. A coarse tuneup of the electric field pattern in the beam channel is achieved by an end resonance element, whereas a fine tuneup is attained by changing the cavity inductance with the angle between the "flaps" (Fig. 1.10). The field strength in the accelerating gaps can be equalized also by stabilizing rods tied to the stems of the drift tubes as in the CO^+ ion accelerator for 153–224 keV/nucleon built at the University of Frankfurt am Main (Fig. 1.11). At 108 MHz, the effective shunt impedance of such an accelerating cavity proved to be rather high—500 MΩ/m [46].

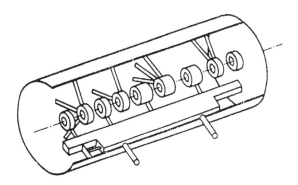

Figure 1.9. Ant-type accelerating structure.

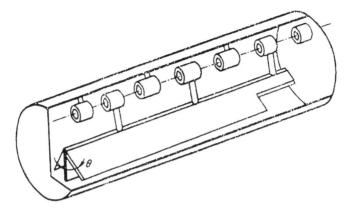

Figure 1.10. Interdigital H-type cavity [45].

Figure 1.11. Interdigital H-type accelerating structure [46].

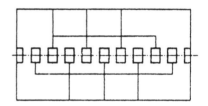

Figure 1.12. Accelerating structure of the Uragan accelerator.

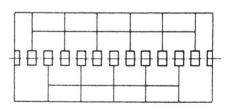

A good performance was demonstrated by a structure resonant in the TE_{111} mode (Fig. 1.12) [47]. On the one hand, it retains a rather high effective shunt impedance characteristic of the previous structure, on the other hand, the field pattern in beam channel is regulated by end resonance elements.

In H-mode accelerating cavities, particles are focused either by an azimuth-independent accelerating field or by an accelerating field having an azimuthal quadrupole symmetry. The

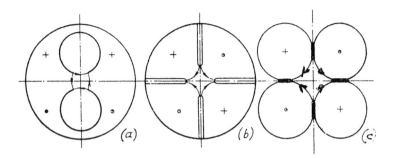

Figure 1.13. (a) Three-chamber and (b, c) four-chamber H-type resonators.

former are called accelerators with phase-variable focusing, the latter are called accelerators with rf quadrupole focusing (RFQ) [9].

Accelerators with RFQ widely use three-chamber and four compartment H-type cavities (Fig. 1.13). The field structure in each half of a three-chamber cavity divided by a virtual central horizontal plane is similar to that in a two-chamber structure. The magnetic field is closed also near the end walls of the cavity short of the chamber partition walls. Accelerating electrodes are fixed to the ends of tube cuts. Similar notions can be done about four-chamber cavities. The diagram at (b) illustrates the geometry of the axially symmetric case. Here, the cavity is a cylinder partitioned by identical radial walls short of the resonator axis. The cavity surface coincides with the coordinate surfaces of the system. The design can be simplified by separation of variables. The diagram at (c) represents the clever leaf cavity.

2

ELECTRODYNAMIC
MODELING OF
ACCELERATING
CAVITIES

2.1 General Background

The object of our investigation in this chapter is resonators, i.e., certain spatial cavities filled with a linear medium whose μ_a and ε_a are independent of the field strength and invariable in time. These cavities are bounded (partially or completely) by a conducting surface S. The microscopic electrodynamics of such cavities may be described by the set of Maxwell's equations and the Lorentz equation for a force acting upon free charges with the corresponding boundary and initial conditions:

$$
\left.
\begin{aligned}
\operatorname{curl} \mathbf{H} &= \mathbf{J} + \varepsilon_a \frac{\partial \mathbf{E}}{\partial t}, \\
\operatorname{curl} \mathbf{E} &= -\mu_a \frac{\partial \mathbf{H}}{\partial t} - \mathbf{J}_m,
\end{aligned}
\right\}
$$

$$\left.\begin{aligned}
&\operatorname{div} \varepsilon_a \mathbf{E} = \rho, \\
&\operatorname{div} \mu_a \mathbf{H} = \rho_m, \\
&\frac{d\mathbf{p}}{dt} = \left(\mathbf{E} + \mu_a \mathbf{v} \times \mathbf{H}\right) \cdot q,
\end{aligned}\right\} \tag{2.1}$$

where $\mathbf{E}$ and $\mathbf{H}$ are the electric and magnetic field strengths, $\mathbf{J}$ is the electric current density, $\mathbf{J}_m$ is the magnetic current density, ρ is the electric charge density, ρ_m is the magnetic charge density, q is the charge, $\varepsilon_a = \varepsilon\varepsilon_0$ is the absolute permittivity of the medium, $\mu_a = \mu\mu_0$ is the permeability of the medium, ε_0 and μ_0 are the permittivity and permeability of vacuum, $\mathbf{v} = d\mathbf{r}/dt$ is the particle velocity, $\mathbf{r}$ is the particle radius-vector, $\mathbf{p} = \gamma m_0 \mathbf{v}$ is the particle momentum, and γ is the particle energy normalized by $m_0 c^2$.

In the general case, equations (2.1) constitute a nonlinear system describing self-consistent processes: the motion of free charges is governed by the fields $\mathbf{E}$ and $\mathbf{H}$ which, in turn, depend on the current density, i.e., on the motion of charges.

One efficient approach to solving (2.1) in cavities consists of dividing the electric and magnetic fields into rotational and potential components, which is equivalent to choosing the Coulomb calibration for the potential. Such a division follows from the Helmholtz expansion theorem stating that any vector field is representable as a sum of two fields [48]: a solenoidal or rotational field whose divergence vanishes and a potential or gradient field whose rotation vanishes. In this representation

$$\mathbf{E}(t) = \mathbf{E}^c(t) + \mathbf{E}^p(t), \quad \mathbf{H}(t) = \mathbf{H}^c(t) + \mathbf{H}^p(t), \tag{2.2}$$

where $\mathbf{E}^c$ and $\mathbf{H}^c$ are the rotational components and $\mathbf{E}^p$ and $\mathbf{H}^p$ are the potential components of the vector fields, so that

$$\left.\begin{aligned}
&\operatorname{div} \mathbf{E}^c \equiv 0, \quad \operatorname{div} \mathbf{H}^c \equiv 0, \\
&\operatorname{curl} \mathbf{E}^p \equiv 0, \quad \operatorname{curl} \mathbf{H}^p \equiv 0.
\end{aligned}\right\} \tag{2.3}$$

The potential component of the magnetic field is usually omitted from the formulation of standard electrodynamic problems because magnetic flux density lines are always closed. However, this is the case for fields in an infinite space. In the finite volume of a cavity bounded by a surface with apertures (such as channels for beams of charged particles or coupling holes), the potential component of the magnetic field does not vanish in the cavity because some magnetic flux density lines leave the cavity and close beyond the volume under consideration. Therefore, the boundary of the cavity can have fragments, from which magnetic flux density lines depart, and fragments, at which they arrive— this pattern is equivalent to the existence of magnetic charges and surface magnetic currents on these fragments.

In agreement with (2.3), the potential components of electromagnetic field can be determined through the scalar potentials

$$\mathbf{E}^{\mathrm{p}} = -\mathrm{grad}\,\varphi, \quad \mathbf{H}^{\mathrm{p}} = -\mathrm{grad}\,\psi. \qquad (2.4)$$

Substituting (2.4) into the corresponding equations of (2.1), we obtain Poisson equations in terms of electric and magnetic potentials:

$$\nabla^2 \varphi = -\frac{\rho}{\varepsilon_a}; \quad \nabla^2 \psi = -\frac{\rho_m}{\mu_a} \qquad (2.5)$$

Note that equations (2.5) are written for a cavity filled with a homogeneous medium whose ε_a and μ_a are independent of spatial coordinates. Thus, at each time instant, the potential components of the electric and magnetic fields are determined as the solution to equations (2.5) with the corresponding boundary conditions and are characterized by the static behavior notwithstanding their dependence on time.

In order to derive equations in terms of the rotational components of electromagnetic fields, we substitute (2.2) into the two first equations of (2.1) and transform the result taking into account (2.3) to obtain

$$\left.\begin{array}{l} \operatorname{curl} \operatorname{curl} \mathbf{H}^c + \varepsilon_a \mu_a \dfrac{\partial^2 \mathbf{H}^c}{\partial t^2} = \operatorname{curl} \mathbf{J} - \varepsilon_a \dfrac{\partial \mathbf{J}_m}{\partial t} - \varepsilon_a \mu_a \dfrac{\partial^2 \mathbf{H}^p}{\partial t^2}; \\[3mm] \operatorname{curl} \operatorname{curl} \mathbf{E}^c + \varepsilon_a \mu_a \dfrac{\partial^2 \mathbf{E}^c}{\partial t^2} = -\operatorname{curl} \mathbf{J}_m - \mu_a \dfrac{\partial \mathbf{J}}{\partial t} - \varepsilon_a \mu_a \dfrac{\partial^2 \mathbf{E}^p}{\partial t^2}. \end{array}\right\} \quad (2.6)$$

These equations demonstrate that, because of the calibration chosen, the rotational components of the electromagnetic fields depend, in the general case, not only on the motion of the electric and magnetic charges, but also on the temporal variations of the corresponding potential components.

Together with the equation for the Lorentz force, equations (2.4)–(2.6) form a new complete set, which describes the processes in cavities and is based on representing the electromagnetic field by a sum of the rotational and potential components.

It is convenient to solve these equations by expanding the unknown fields in series in the eigenfunctions of the corresponding differential operators [49].

A complete set of scalar eigenfunctions for the Laplace operator in equation (2.5) is obtained by solving the scalar Helmholtz equations

$$\nabla^2 \varphi + k^2 \varphi = 0, \quad \nabla^2 \psi + k^2 \psi = 0 \qquad (2.7)$$

within the cavity with the homogeneous boundary conditions $\varphi = 0$, $\partial \psi / \partial \mathbf{n} = 0$, which correspond to the boundary conditions for the electric and magnetic fields

$$\mathbf{n} \times \mathbf{E}\big|_S = 0; \quad \mathbf{n} \cdot \mathbf{H}\big|_S = 0 \qquad (2.8)$$

on the boundary of the cavity. Here, $\mathbf{n}$ is the external normal to the surface of the cavity.

The complete set of scalar eigenfunctions for the Laplace operator is then used to construct the complete set of vector potentials $\{\mathbf{E}_g, \mathbf{H}_g\}$, where the vector functions $\mathbf{E}_g$ and $\mathbf{H}_g$ satisfy the equations

$$\mathbf{E}_g = -\operatorname{grad}\varphi_g, \quad \mathbf{H}_g = -\operatorname{grad}\psi_g, \tag{2.9}$$

the orthonormalization conditions

$$\int_V \mathbf{E}_g \mathbf{E}_j dv = \delta_{gj}, \quad \int_V \mathbf{H}_g \mathbf{H}_j dv = \delta_{gj}, \tag{2.10}$$

(δ_{gi} is the Kronecker delta), and homogeneous boundary conditions (2.8). In this case, functions ψ_g and φ_g satisfy the equations

$$\nabla^2 \psi_g + k^2 \psi_g = 0,$$

$$\nabla^2 \varphi_g + k^2 \varphi_g = 0.$$

Therefore, the potential electromagnetic components, which are the solution to equations (2.5), are representable as the series

$$\mathbf{E}^p = \sum_g V_g \mathbf{E}_g, \quad \mathbf{H}^p = \sum_g I_g \mathbf{H}_g. \tag{2.11}$$

The complete set of vector eigenfunctions for the differential operator in equations (2.6) is obtained by solving the homogeneous differential equations in the cavity for the case of the harmonic temporal behavior of the vectors

$$\operatorname{curl}\operatorname{curl}\mathbf{H}^c - k^2\mathbf{H}^c = 0,$$

$$\operatorname{curl}\operatorname{curl}\mathbf{E}^c - k^2\mathbf{E}^c = 0. \tag{2.12}$$

Then, the rotational components of the electric and magnetic fields satisfying equations (2.6) can be represented as the expansions

$$\mathbf{E}^c(t) = \sum_m V_m(t)\mathbf{E}_m, \quad \mathbf{H}^c(t) = \sum_m I_m(t)\mathbf{H}_m, \tag{2.13}$$

where the vector functions $\mathbf{E}_m$ and $\mathbf{H}_m$ satisfy the relationships

$$\operatorname{curl}\mathbf{H}_m = k_m\mathbf{E}_m, \quad \operatorname{curl}\mathbf{E}_m = k_m\mathbf{H}_m \qquad (2.14)$$

and, consequently, equations (2.12), the orthonormalization conditions

$$\int\limits_V \mathbf{E}_m\mathbf{E}_j dv = \delta_{mj}, \quad \int\limits_V \mathbf{H}_m\mathbf{H}_j dv = \delta_{mj}, \qquad (2.15)$$

and the homogeneous boundary conditions (2.8).

Thus, the solution of combined sets (2.6) and (2.5) can be found in two steps: first equations (2.7) and (2.12) are solved to find the scalar and vector eigenfunctions for the corresponding differential operators and then determine the unknown coefficients V_g, I_g, V_m, and I_m in the expansion of the electromagnetic field. This approach is not uniformly efficient, but it has certain advantages for many practical applications, specifically for stationary processes in cavities.

The solution to equations (2.12) is valuable by itself because practically important characteristics of cavities, such as the resonance frequencies (resonance wavenumbers) and the spatial distribution of the electromagnetic field, very closely correspond to the eigenvalues and eigenfunctions obtained from the solution of (2.12).

Even in the case of homogeneous boundary conditions the analytical solution to (2.12) is possible only for a few cases, particularly if the surface bounding the cavity is a coordinate surface, so that the Fourier separation of variables is applicable. Among the cavities bounded by coordinate surfaces we note cylindrical resonators (in a cylindrical coordinate system), rectangular prismatic resonators (in a Cartesian coordinate system), coaxial resonators (in a cylindrical coordinate system), spherical resonators (in a spherical coordinate system), and so on. In practice, cavities shaped close to cylindrical and rectangular prismatic resonators are the most frequent designs. Applying the Fourier method to solve equations (2.12) for these types of resonators subject to the homogeneous boundary conditions (2.8), we single out two groups of solutions, which differ in the component structure of the electric and magnetic fields:

- TM-mode (E-mode) characterized by a non-zero longitudinal component of the electric field and by the purely transverse magnetic field;
- TE-mode (H-mode) characterized by a non-zero longitudinal component of the magnetic field and by the purely transverse electric field;

In this representation, the rotational eigenfunctions of the rectangular resonator can be written as follows:

for TM cavity modes,

$$
\left.
\begin{aligned}
E_x &= A_p \frac{mp\pi^2}{aL} \frac{1}{k_e} \frac{1}{\kappa_e} \cos\frac{m\pi x}{a} \sin\frac{n\pi y}{b} \sin\frac{p\pi z}{L}, \\[2mm]
E_y &= A_p \frac{np\pi^2}{bL} \frac{1}{k_e} \frac{1}{\kappa_e} \sin\frac{m\pi x}{a} \cos\frac{n\pi y}{b} \sin\frac{p\pi z}{L}, \\[2mm]
E_z &= A_p \frac{\kappa_e}{k_e} \sin\frac{m\pi x}{a} \sin\frac{n\pi y}{b} \cos\frac{p\pi z}{L}, \\[2mm]
H_x &= -A_p \frac{1}{\kappa_e} \frac{n\pi}{b} \sin\frac{m\pi x}{a} \cos\frac{n\pi y}{b} \cos\frac{p\pi z}{L}, \\[2mm]
H_y &= A_p \frac{1}{\kappa_e} \frac{m\pi}{a} \cos\frac{m\pi x}{a} \sin\frac{n\pi y}{b} \cos\frac{p\pi z}{L}, \\[2mm]
H_z &= 0,
\end{aligned}
\right\} \tag{2.16}
$$

and for TE cavity modes,

$$
\begin{aligned}
E_x &= -B_{mn} \frac{1}{\kappa_h} \frac{n\pi}{b} \cos\frac{m\pi x}{a} \sin\frac{n\pi y}{b} \sin\frac{p\pi z}{L}, \\[2mm]
E_y &= B_{mn} \frac{1}{\kappa_h} \frac{m\pi}{a} \sin\frac{m\pi x}{a} \cos\frac{n\pi y}{b} \sin\frac{p\pi z}{L}, \\[2mm]
E_z &= 0,
\end{aligned}
$$

$$H_x = -B_{mn} \frac{mp\pi^2}{aL} \frac{1}{k_h} \frac{1}{\kappa_h} \sin\frac{m\pi x}{a} \cos\frac{n\pi y}{b} \cos\frac{p\pi z}{L},$$

$$H_y = -B_{mn} \frac{np\pi^2}{bL} \frac{1}{k_h} \frac{1}{\kappa_h} \cos\frac{m\pi x}{a} \sin\frac{n\pi y}{b} \cos\frac{p\pi z}{L},$$

$$H_z = B_{mn} \frac{\kappa_h}{k_h} \cos\frac{m\pi x}{a} \cos\frac{n\pi y}{b} \sin\frac{p\pi z}{L},$$

$$(2.17)$$

where

$$A_p = 2\sqrt{\frac{2-\delta_{0p}}{abL}}, \quad B_{mn} = \sqrt{\frac{2(2-\delta_{0m})(2-\delta_{0n})}{abL}},$$

ab is the cavity cross section; L is the cavity length (the origin of the coordinate system is placed at a prism vertex);

$$\kappa_{h,e} = \sqrt{(m\pi/a)^2 + (n\pi/b)^2}$$

are the transverse wavenumbers (e for TM-modes, h for TE-modes);

$$k_{h,e} = \sqrt{\kappa_{h,e}^2 + (p\pi/L)^2} = 2\pi f_{h,e}/c$$

are the resonance wavenumbers; c is the velocity of light in vacuum; m, n, and p are the numbers of half periods of the wave along x, y, and z coordinates, respectively; and δ_{0j} is the Kronecker delta.

Similarly, for a cylindrical cavity, the rotational vector eigenfunctions can be written as follows:

for the TM cavity modes,

$$E_r = \frac{A}{k_e R} \frac{p\pi}{L} \frac{J_{m-1}(\nu_{mn}\, r/R) - J_{m+1}(\nu_{mn}\, r/R)}{2J_{m-1}(\nu_{mn})}$$

$$\times \cos(m\varphi + \pi)\sin\frac{p\pi z}{L},$$

$$E_\varphi = -\frac{A}{k_e r} \frac{m}{v_{mn}} \frac{p\pi}{L} \frac{J_m(v_{mn} r/R)}{J_{m-1}(v_{mn})} \times \sin(m\varphi + \pi)\sin\frac{p\pi z}{L},$$

$$E_z = -\frac{A v_{mn}}{k_e R^2} \frac{J_m(v_{mn} r/R)}{J_{m-1}(v_{mn})} \times \cos(m\varphi + \pi)\cos\frac{p\pi z}{L},$$

$$H_r = \frac{A}{r} \frac{m}{v_{mn}} \frac{J_m(v_{mn} r/R)}{J_{m-1}(v_{mn})} \times \sin(m\varphi + \pi)\cos\frac{p\pi z}{L},$$

$$H_\varphi = \frac{A}{r} \frac{J_{m-1}(v_{mn} r/R) - J_{m+1}(v_{mn} r/R)}{2J_{m-1}(v_{mn})}$$

$$\times \cos(m\varphi + \pi)\cos\frac{p\pi z}{L},$$

$$H_z = 0,$$

(2.18)

and, for the TE cavity modes,

$$E_r = -\frac{B}{r} \frac{m}{\sqrt{\mu_{mn}^2 - m^2}} \frac{J_m(\mu_{mn} r/R)}{J_m(\mu_{mn})}$$

$$\times \sin(m\varphi + \pi)\sin\frac{p\pi z}{L},$$

$$E_\varphi = -\frac{B}{R} \frac{\mu_{mn}}{\sqrt{\mu_{mn}^2 - m^2}} \frac{J_{m-1}(\mu_{mn} r/R) - J_{m+1}(\mu_{mn} r/R)}{2J_m(\mu_{mn})}$$

$$\times \cos(m\varphi + \pi)\sin\frac{p\pi z}{L},$$

$$E_z = 0,$$

$$H_r = \frac{B}{k_h R} \frac{\mu_{mn}}{\sqrt{\mu_{mn}^2 - m^2}} \frac{p\pi}{L} \frac{J_{m-1}(\mu_{mn} r/R) - J_{m+1}(\mu_{mn} r/R)}{2J_m(\mu_{mn})}$$

(2.19)

$$\times \cos(m\varphi + \pi)\cos\frac{p\pi z}{L},$$

$$H_\varphi = -\frac{B}{k_h r}\frac{m}{\sqrt{\mu_{mn}^2 - m^2}}\frac{p\pi}{L}\frac{J_m(\mu_{mn}\,r/R)}{J_m(\mu_{mn})}$$

$$\times \sin(m\varphi + \pi)\cos\frac{p\pi z}{L},$$

$$H_z = \frac{B}{k_h}\frac{\mu_{mn}^2}{R^2}\frac{1}{\sqrt{\mu_{mn}^2 - m^2}}\frac{J_m(\mu_{mn}\,r/R)}{J_m(\mu_{mn})}$$

$$\times \cos(m\varphi + \pi)\sin\frac{p\pi z}{L},$$

where R is the radius of the cavity, L is the length of the cavity,

$$k_e = \sqrt{(v_{mn}/R)^2 + (p\pi/L)^2},$$

$$k_h = \sqrt{(\mu_{mn}/R)^2 + (p\pi/L)^2}$$

are the resonance numbers, v_{mn} is the nth root of the Bessel function J_m, μ_{mn} is the nth root of the first derivative of the Bessel function J_m, m indicates the number of full circumferential periods, n is the number of radial variation, and p is the number of half periods along the z axis.

The rotational vector eigenfunctions of rectangular prismatic and cylindrical cavities, given in a component representation in (2.16)–(2.19), are orthogonal and normalized:

$$\int_V E_m^2 dv = 1, \quad \int_V H_m^2 dv = 1. \tag{2.20}$$

Here, m is the ordinal number of the rotational function.

For cavities bounded by a coordinate surface, the Fourier method can be used also for solving the scalar equations (2.7) subject to the homogeneous boundary conditions (2.8) to construct a set of potential vector eigenfunctions $\{E_g, H_g\}$. Solving equations (2.7) in the rectangular prismatic and cylindrical cavities, one can write the corresponding potential eigenfunctions as [49]

$$\mathbf{E}_g = \sqrt{\frac{2}{L}} \frac{1}{k_g} \left(\nabla \psi_e \sin\frac{p\pi z}{L} + \psi_e \mathbf{z}\frac{p\pi}{L}\cos\frac{p\pi z}{L} \right);$$

$$\mathbf{H}_g = \sqrt{\frac{2-\delta_{0p}}{L}} \frac{1}{k_g} \left(\nabla \psi_h \cos\frac{p\pi z}{L} - \psi_h \mathbf{z}_0\frac{p\pi}{L}\sin\frac{p\pi z}{L} \right),$$

$$\left. \right\} \quad (2.21)$$

where

$$k_g = \sqrt{\kappa_{h,e}^2 - (p\,\pi/L)^2}$$

are the eigenvalues of the problem, and scalar functions $\psi_{e,h}$ and constants $\kappa_{h,e}$ are determined as follows:

for rectangular cavities,

$$\psi_e = \frac{2}{\sqrt{ab}} \sin\frac{m\pi x}{a} \sin\frac{n\pi y}{b};$$

$$\psi_h = \sqrt{\frac{(2-\delta_{0m})(2-\delta_{0n})}{ab}} \cos\frac{m\pi x}{a} \cos\frac{n\pi y}{b};$$

$$\kappa_{h,e} = \sqrt{\left(\frac{m\pi}{a}\right)^2 + \left(\frac{n\pi}{b}\right)^2},$$

$$\left. \right\} \quad (2.22)$$

and for cylindrical cavities

$$\psi_e = \sqrt{\frac{2-\delta_{0m}}{\pi}} \frac{1}{R} \frac{J_m(\nu_{mn}\,r/R)}{J_{m-1}(\nu_{mn})}\cos m\varphi;$$

$$\psi_h = \sqrt{\frac{2-\delta_{0m}}{\pi}} \frac{1}{R} \frac{\mu_{mn}}{\sqrt{\mu_{mn}^2 - m^2}} \frac{J_m(\mu_{mn}\,r/R)}{J_m(\mu_{mn})}\cos m\varphi;$$

$$\kappa_e = \nu_{mn}/R;$$

$$\kappa_h = \mu_{mn}/R.$$

$$\left. \right\} \quad (2.23)$$

Here, we used the same notation as in (2.16)–(2.19).

Like rotational eigenfunctions, potential eigenfunctions (2.21) are orthogonal and normalized:

$$\int_V \mathbf{E}_g^2 dv = 1, \quad \int_V \mathbf{H}_g^2 dv = 1 . \tag{2.24}$$

The Fourier method gives us a possibility of analytically constructing sets of vector eigenfunctions for other cavities bounded by coordinate surfaces. However, most cavities used in practice have more intricate geometry, which complicates analytical solutions to equations (2.12) and (2.7). For such cavities, a set of vector eigenfunctions can be constructed only by numerical methods based on one or another discretization procedure of the initial equations. Obviously, this approach has some disadvantages to it. One of them is that only relatively small number of vector eigenfunctions can be actually evaluated. This circumstance limits the area of solvable problems mainly to stationary and slow transient processes, which do not require a significant number of terms in series (2.11) and (2.13).

2.2 Numerical Procedures and RF Cavity Design Codes

In Sect. 2.1, we demonstrated the practical importance of the solution to equations (2.12) because it yields important characteristics of cavities, such as resonance frequencies and the corresponding spatial pattern of the electric and magnetic fields. This circumstance lent impetus to the development of numerical procedures and the corresponding computer codes for evaluating the eigenfunctions and eigenvalues of complex-geometry cavities used in charged-particle accelerators and microwave instruments. Instead of equations (2.12), many authors use as a point of departure the from differential equations

$$\nabla^2 \mathbf{E} + k^2 \mathbf{E} = 0,$$
$$\nabla^2 \mathbf{H} + k^2 \mathbf{H} = 0, \tag{2.25}$$

which are a direct consequence of (2.12). Both sets are equivalent in their rotational vector eigenfunctions, but differ in numerical analysis and codes.

In the general case, the numerical analysis of equation (2.12) [or (2.25)] is a complex problem, which can be considered as a number of problems of increasing complexity:

step 1 seeks the characteristics of rf cavities with cylindrical symmetry for the lowest mode. This formulation is equivalent to the two-dimensional scalar problem for the azimuthal component of the magnetic field (for TM modes) or electric field (for TE modes) in a longitudinal section of the cavity;

step 2 seeks the characteristics of rf cavities with cylindrical symmetry for azimuth-independent modes. This formulation is also equivalent to the two-dimensional scalar problem for the azimuthal component of the magnetic or electric field in a longitudinal section of the cavity, but requires higher eigenvalues and the corresponding eigenfunctions;

step 3 seeks the characteristics of rf cavities with cylindrical symmetry for azimuth-dependent modes. This formulation is equivalent to the two-dimensional vector problem in a longitudinal section of the cavity;

step 4 seeks the characteristics of asymmetric rf cavities for various oscillation modes. This formulation is equivalent to the three-dimensional vector problem in the volume of the cavity.

The majority of numerical procedures for the partial differential equations (2.12) and (2.25) use certain discretization techniques to reduce them to a set of linear algebraic equations. In practice, four basic methods are used to solve boundary-value problems in cavities: the finite difference method, the variation method, the integral equation method, and the finite element analysis that is widely used lately. Because discretization can be viewed as a projection of an infinite-dimensional functional space on a finite-dimensional space, all the above methods are versions of the projection method. They have been considered in detail by a number of authors [50–59].

One of the most efficient numerical methods for solving equations (2.12) and (2.25) is the finite difference method, which offers simple conversation into computer codes. Using this method, researchers of the Los-Alamos National Laboratory designed the first program for evaluating the characteristics of rf cavities with cylindrical symmetry for the lowest mode in 1966. This program, known as LALA [60], was widely used in accelerator design at that time. Although this code was not so adaptable to different geometries of cavities as modern programs, it provided reliable and sufficiently accurate results and was a good instrument for numerical analysis of many cavities. The LALA program used the successive upper relaxation to find the azimuthal component of magnetic field from the set of linear algebraic equations produced by the quantization of the initial Helmholtz equation. This method ensures a high convergence for the smallest eigenvalue (the resonance frequency of the lowest mode).

The LALA code was a progenitor of more sophisticated modern programs based on the finite difference method. These programs, collectively called GNOM, solve rf cavities with cylindrical symmetry for both azimuth-independent and azimuth-dependent modes [62–66]. The GNOM program package has a modular structure, is adaptable to different cavity geometries, runs in a dialog mode, and provides reliable results.

Among other modern Russian programs using the finite difference method for the quantization of the initial equations, we point out the AZIMUTH program [67–68] designed at the Institute of Electrical Engineering, St. Petersburg, to solve rf cavities with cylindrical symmetry for azimuth-independent and azimuth-dependent oscillation modes. The AZIMUTH program is very fast and stable. Its relative calculation error in resonance frequencies of azimuth-dependent oscillation modes decreases when the azimuth field number increases. A disadvantage of this program is that the calculation time essentially depends on the ordinal number of the calculated oscillation mode because it involves a preliminary calculation of all other modes having the same azimuth field number, but smaller resonance frequencies. For this reason, the potential of this code is mainly limited by evaluating resonance frequencies of several first modes with a fixed azimuth field number.

One of top-level finite difference programs for evaluating azimuth dependent oscillation modes in cylindrically symmetric rf cavities is the URMEL code [69] designed in Germany. This program solves the set of linear algebraic equations produced by the digitization of the initial equations in the radial and longitudinal components of the magnetic field using the backward iteration scheme.

The method of finite differences also lies at the basis of the CAV3D code [70], which is the first program that actually solves three-dimensional boundary-value problems in cavities. For the quantization of the initial equations (2.25), the program approximates partial derivatives using a seven-point scheme on a regular cubic mesh. The CAV3D program solves the boundary-value problem using the simultaneous forward iteration procedure in which every iteration step requires the calculation of several products of a sparse matrix and a vector. To reject nonphysical solutions, the program uses a penalty method: the initial differential equation (2.25) is augmented with the penalty term

$$\left[\nabla^2 + k^2 + \nu \operatorname{grad} \operatorname{div}\right] \mathbf{H} = 0 \qquad (2.26)$$

and the problem is solved several times with different ν. Solutions dependent on ν are rejected. The main disadvantage of the CAV3D program is that the relative calculation error in resonance frequencies is large.

Another efficient method that is widely used for solving various problems of mathematical physics is the finite element analysis. Among the first known programs based on the finite element method we point out the SUPERFISH program [71] whose appearance in 1976 has given impetus to the design of many improved programs.

The SUPERFISH program is intended for designing rf cavities with cylindrical symmetry which operate at azimuth-independent modes. It uses finite elements on an irregular triangular mesh, which allows a very accurate approximation of the cavity boundary. In order to determine eigenvalues, the program seeks for the zero of the residual function of the boundary-value problem with a circular exciting magnetic current at one of the mesh nodes. Unfortunately, this program is prone to problems related to the correct choice of a node with circular exciting current because this choice is governed by both the cavity geometry and the spatial pattern of the electromagnetic field of the desired mode, and this dual dependence impedes the conversion of the choice to an algorithm.

A modified finite element calculation of the eigenfunctions and resonance frequencies of cavities has been used in the programs LANS [72] and LANS2 [73] for azimuth-independent and azimuth-dependent oscillation modes respectively. Unlike the SUPERFISH program, these programs use backward iterations with a frequency shift to solve the set of equations obtained by the quantization of the initial equations. It is assumed that the cavity is excited by magnetic currents distributed over the resonator longitudinal section rather than by a single circular magnetic current. A disadvantage of these programs is an increased (over SUPERFISH) memory requirement for temporary storage.

One more modern finite element code is the PRUD program package [74, 75] solving rf cavities with cylindrical symmetry for azimuth-dependent modes. A similar recent program is ULTRAFISH [76]. Both programs have a common essential disadvantage: the matrix of the set contains singular elements, which impedes working with the programs and often increases the calculation error.

The finite element method is used also in the fast solvers MULTIMODE [77] and PRUD-0 [78] for rf cavities with cylindrical symmetry for various modes. These programs solve the algebraic eigenvalue problem by iterations in a subspace. In contrast to previous programs, the MULTIMODE and PRUD-0 use ultraparametric quadrangular finite elements to enhance the calculation accuracy and efficiency.

Early programs using the finite element method for solving three-dimensional boundary-value problems in cavities [79–

81] have been refined later. The most versatile and sophisticated program package of this type is the MAFIA CAD-system [82, 83].

The integral equation method is seldom used for solving electrodynamic boundary-value problems in cavities. Among known programs based on the integral equation method and intended for evaluating oscillation modes in rf cavities with cylindrical symmetry, we point out the MAXWELL code [84] which reduces the problem of determining the fields in a cavity to finding the characteristics of surface field sources. This program uses the finite element method to quantize the equation in surface distribution of the tangential magnetic field, and solves the algebraic eigenvalue problem by backward iterations. In cylindrical cavities, its calculation error in resonance frequencies for azimuth-dependent oscillation modes is about 10^{-2}–10^{-3}.

2.3   Electrodynamic Models of Coupled Cavities

Consider a system of N coupled cavities. For each cavity, we can write Maxwell's equations (2.1). We consider the coupled cavity system under the condition that currents and charges are absent. In addition, we will not take into account the field asymmetry produced by intercavity coupling elements. Then, we can write the system (2.1) for the pth cavity in the form

$$\left.\begin{aligned}
\operatorname{curl} \mathbf{H}_p &= \varepsilon_a \frac{\partial \mathbf{E}_p}{\partial t}, \\[2mm]
\operatorname{curl} \mathbf{E}_p &= -\mu_a \frac{\partial \mathbf{H}_p}{\partial t}, \\[2mm]
\operatorname{div} \varepsilon_a \mathbf{E}_p &= 0, \\[2mm]
\operatorname{div} \mu_a \mathbf{H}_p &= 0,
\end{aligned}\right\} \tag{2.27}$$

where $p = 1, 2, 3, \ldots, N$.

We solve the set (2.27) taking into account only rotational fields. The boundary conditions (2.8) are as before, i.e.,

$$\mathbf{n} \times \mathbf{E}_p\big|_S = 0, \quad \mathbf{n} \cdot \mathbf{H}_p\big|_S = 0 . \tag{2.28}$$

According to expression (2.13), we write the rotational field in a cell as a series in the orthogonal eigenfunctions

$$\left.\begin{aligned} \mathbf{E}_p(t) &= \sum_m V_{pm}\mathbf{E}_{pm}(t), \\ \mathbf{H}_p(t) &= \sum_m I_{pm}\mathbf{H}_{pm}(t), \end{aligned}\right\} \tag{2.29}$$

where m is the number of modes taken into consideration; V_{pm} and I_{pm} are the decomposition factors for the electric and magnetic fields, respectively; and $\mathbf{E}_{pm}$ and $\mathbf{H}_{pm}$ are the eigenfunctions of the pth cavity satisfying the equations

$$\left.\begin{aligned} \operatorname{curl} \mathbf{H}_{pm} &= k_{pm}\mathbf{E}_{pm}, \\ \operatorname{curl} \mathbf{E}_{pm} &= k_{pm}\mathbf{H}_{pm}, \\ \operatorname{div} \mathbf{E}_{pm} &= \operatorname{div} \mathbf{H}_{pm} = 0, \end{aligned}\right\} \tag{2.30}$$

the boundary conditions (2.28), and the orthonormalization conditions

$$\left.\begin{aligned} \int_V \mathbf{E}_{pi} \cdot \mathbf{E}_{pj} dv &= \delta_{ij}, \\ \int_V \mathbf{H}_{pi} \cdot \mathbf{H}_{pj} dv &= \delta_{ij} \end{aligned}\right, \tag{2.31}$$

where δ_{ij} is the Kronecker delta and k_{pm} is the wavenumber of the mth oscillation mode in the pth cavity.

For harmonic waves

$$\mathbf{E}_p(t) = \mathbf{E}_p e^{j\omega t}, \quad \mathbf{H}_p(t) = \mathbf{H}_p e^{j\omega t}, \tag{2.32}$$

Maxwell's equations become

$$\begin{aligned} \operatorname{curl} \mathbf{H}_p &= j\omega\varepsilon_a \mathbf{E}_p, \\ \operatorname{curl} \mathbf{E}_p &= -j\omega\mu_a \mathbf{H}_p. \end{aligned} \tag{2.33}$$

Combining equations (2.28)–(2.33) and using vector calculus, we obtain equations for the unknown decomposition factors of the electric (V_{pm}) and magnetic (I_{pm}) fields.

Multiplying the first equation in (2.33) by $\mathbf{E}_{pm}$ and integrating the result over the cavity volume, we have

$$\int_V \mathbf{E}_{pm} \cdot \operatorname{curl} \mathbf{H}_p \, dv = j\omega\varepsilon_a \int_V \mathbf{E}_{pm} \cdot \mathbf{E}_p \, dv. \tag{2.34}$$

Using some vector calculus formulas, we rearrange the expression in the left-hand side of (2.34) to obtain

$$\int_V \mathbf{E}_{pm} \cdot \operatorname{curl} \mathbf{H}_p \, dv = \int_V \mathbf{H}_p \cdot \operatorname{curl} \mathbf{E}_{pm} \, dv - \int_V \operatorname{div}(\mathbf{E}_{pm} \times \mathbf{H}_p) \, dv$$

$$= k_{pm} \int_V \mathbf{H}_p \cdot \mathbf{H}_{pm} \, dv - \oint_S (\mathbf{E}_{pm} \times \mathbf{H}_p) \cdot \mathbf{n} \, ds$$

$$= k_{pm} I_{pm} - \oint_S (\mathbf{n} \times \mathbf{E}_{pm}) \cdot \mathbf{H}_p \, ds = k_{pm} I_{pm}. \tag{2.35}$$

Taking into account the orthonormalization condition (2.31), we find

$$k_{pm} I_{pm} = j\omega\varepsilon_a V_{pm}. \tag{2.36}$$

Similarly, multiplying the second equation in (2.33) by $\mathbf{H}_{pm}$ and integrating the result over the volume V, we obtain

$$k_{pm}V_{pm} + \oint_S (\mathbf{E}_p \times \mathbf{H}_{pm}) \cdot \mathbf{n}\, ds = -j\omega\mu_a I_{pm} .$$ (2.37)

Combining equations (2.36) and (2.37), we find the set of equations in the unknown decomposition factors of the electric and magnetic fields

$$\left.\begin{aligned}
\left(k_{pm}^2 - k^2\right)V_{pm} + k_{pm} \oint_S (\mathbf{E}_p \times \mathbf{H}_{pm}) \cdot \mathbf{n}\, ds = 0; \\
\left(k_{pm}^2 - k^2\right)I_{pm} + j\omega\varepsilon_a \oint_S (\mathbf{E}_p \times \mathbf{H}_{pm}) \cdot \mathbf{n}\, ds = 0.
\end{aligned}\right\}$$ (2.38)

Consider the integral over the cavity surface S. We represent this integral as

$$\oint_S (\mathbf{E}_p \times \mathbf{H}_{pm}) \cdot \mathbf{n}\, ds = \oint_{S_w} (\mathbf{E}_p \times \mathbf{H}_{pm}) \cdot \mathbf{n}\, ds + \oint_{S_s} (\mathbf{E}_p \times \mathbf{H}_{pm}) \cdot \mathbf{n}\, ds ,$$ (2.39)

where S_s is the surface of the coupling aperture and S_w is the metal cavity boundary.

The integral of the vector product of the electric field and the eigenfunction of the magnetic field over the metal surface characterizes the loss of the rf power in the cavity walls and is a factor limiting the decomposition factors I_{pm} and V_{pm} in the absence of coupling apertures.

According to the Shchukin–Rykov–Leontovich boundary conditions, on a metallic surface, the tangential components of the electric and magnetic fields are related by the expression

$$\mathbf{E}_\tau = Z_S \mathbf{H}_\tau \times \mathbf{n} ,$$ (2.40)

where

$$Z_S = (1+j)\sqrt{\mu_a\omega/(2\sigma)}$$

is the surface impedance.

Because on the boundary

$$\mathbf{E}_\tau = (\mathbf{n} \times \mathbf{E}) \times \mathbf{n} , \qquad (2.41)$$

the boundary condition (2.40) reduces to

$$\mathbf{n} \times \mathbf{E} = Z_S \, \mathbf{H} . \qquad (2.42)$$

With this condition we obtain

$$\oint_{S_w} (\mathbf{E}_p \times \mathbf{H}_{pm}) \cdot \mathbf{n} ds = \mu_0 \omega (1+j) \frac{\delta}{2} \oint_{S_w} \mathbf{H}_p \cdot \mathbf{H}_{pm} \, ds .$$

In order to obtain a particular expression for the integral over the inner surface of a cavity, we must represent $\mathbf{H}_p$ as a series in an appropriate functional basis. Therefore, a consideration of the finite conductivity of the cavity walls is sufficient to obtain a set of linearly independent equations even if the cavity has no apertures on its shell. For the sake of simplicity, we assume that the coupling between different modes is small. Taking into account (2.29), we have

$$\oint_{S_w} (\mathbf{E}_p \times \mathbf{H}_{pm}) \cdot \mathbf{n} ds = I_{pm} \mu_a \omega (1+j) \frac{\delta}{2} \oint_{S_w} H_{pm}^2 \, ds , \qquad (2.43)$$

Representing the quality factor of the pth cavity in the mth oscillation mode as

$$Q_{pm} = \frac{2}{\delta} \frac{1}{\oint_{S_w} H_{pm}^2 ds} ,$$

we obtain

$$\oint_{S_w} (\mathbf{E}_p \times \mathbf{H}_{pm}) \cdot \mathbf{n} ds = \frac{(1+j) \mu_a \omega I_{pm}}{Q_{pm}} . \qquad (2.44)$$

Using (2.24), (2.36), and (2.39), we transform the set (2.38) in the unknown decomposition factors for the electric and magnetic fields to the form

$$
\left.\begin{aligned}
\left[k_{pm}^2 - k^2\left(1 - \frac{j-1}{Q_{pm}}\right)\right]V_{pm} + k_{pm}\oint_{S_s}(\mathbf{E}_p \times \mathbf{H}_{pm}) \cdot \mathbf{n}\,ds = 0, \\[2ex]
\left[k_{pm}^2 - k^2\left(1 - \frac{j-1}{Q_{pm}}\right)\right]I_{pm} + j\omega\varepsilon_a\oint_{S_s}(\mathbf{E}_p \times \mathbf{H}_{pm}) \cdot \mathbf{n}\,ds = 0.
\end{aligned}\right\} \quad (2.45)
$$

A further transformation of this set of equations requires considering the integral over the coupling aperture. The concepts of modeling of intercavity coupling elements are discussed in detail in the following chapter using on-axis coupled structure as an example. Here, we note two popular models in determining the coupling between cavities. One presents a coupling slot as a short-circuited segment of a transmission line and another represents a coupling aperture as a waveguide segment aligned with the longitudinal axis of the system. The field excited in a coupling element is determined as a series in the waveguide eigenfunctions satisfying the wave equation and the boundary conditions $\mathbf{E} \times \mathbf{n} = 0$ and $\mathbf{H} \cdot \mathbf{n} = 0$ on the side walls of the waveguide and $\mathbf{E} \cdot \mathbf{n} = 0$ and $\mathbf{H} \times \mathbf{n} = 0$ on its end walls. This model of coupling elements is more general than the representation of coupling slots as a lossless transmission line and handles coupling elements of various shapes. In both cases, the tangential electric field on coupling apertures is determined through the tangential magnetic fields in the cells connected by the coupling aperture.

A coupling slot can be represented as a rectangular cavity with longitudinal size slightly exceeding the wall thickness. In this case, the electromagnetic field in the slot is a superposition of the eigenfunctions of this rectangular resonator [64]:

$$
\mathbf{E}_{sl} = \sum_l V_{sl\,l}\mathbf{E}_{sl\,l}, \quad \mathbf{H}_{sl} = \sum_l I_{sl\,l}\mathbf{H}_{sl\,l}. \quad (2.46)
$$

Using the procedure described above, we obtain expressions for the unknown $V_{\mathrm{sl}\,l}$ and $I_{\mathrm{sl}\,l}$

$$\left.\begin{aligned} \left[k_{\mathrm{sl}\,l}^2 - k^2\left(1 - \frac{j-1}{Q_{\mathrm{sl}\,l}}\right)\right]V_{\mathrm{sl}\,l} + k_{\mathrm{sl}\,l}\oint_{\sum S_i}(\mathbf{E}_m \times \mathbf{H}_{ml})\cdot\mathbf{n}\,ds = 0; \\[2mm] \left[k_{\mathrm{sl}\,l}^2 - k^2\left(1 - \frac{j-1}{Q_{\mathrm{sl}\,l}}\right)\right]I_{\mathrm{sl}\,l} + j\omega\varepsilon_a\oint_{\sum S_i}(\mathbf{E}_m \times \mathbf{H}_{ml})\cdot\mathbf{n}\,ds = 0, \end{aligned}\right\} \quad (2.47)$$

where $\sum S_i = S_1 + S_2 + S_1' + S_2'$

To obtain equations in the unknown decomposition factors of the electromagnetic field in the cavity (V_{pm} and I_{pm}), we resort to the continuity conditions for the electric and magnetic fields at the interface between the cavity and the coupling slot.

At a small distance from the coupling slot, we set a plane as a conditional boundary between the cavity and the slot. This setting is dictated by the fact that right at the slot the tangential components of the electric fields $\mathbf{E}_{pm}$ vanish because they are the orthonormal eigenfunctions for the cavity without slots.

For the energy flux through the conditional boundary between the cavity and the coupling slot to correspond the energy flux through the slot itself, we require that

$$S_1 \gg S_1' \quad \text{and} \quad S_2 \gg S_2', \qquad (2.48)$$

i.e., that the slot areas S_1 and S_2 at interfaces with cavities p and $(p + 1)$ were larger than the lateral areas S_1' and S_2' of the conditional resonators of the coupling slot in these cavities.

At the conditional boundary S_1, the continuity condition for the tangential electric field is [48]

$$\mathbf{n} \times \mathbf{E}_p = \mathbf{n} \times \mathbf{E}_{\mathrm{sl}}. \qquad (2.49)$$

We multiply (2.49) by $\mathbf{H}_{pm}$ and integrate the result over S_1:

$$\int_{S_1} \mathbf{H}_{pm}\cdot(\mathbf{n} \times \mathbf{E}_p)\,ds = \oint_{S_1} \mathbf{H}_{pm}\cdot(\mathbf{n} \times \mathbf{E}_{\mathrm{sl}})\,ds. \qquad (2.50)$$

Using vector calculus formulae and (2.46), we obtain

$$\oint_{S_1}(\mathbf{E}_p \times \mathbf{H}_{pm}) \cdot \mathbf{n}ds = \sum_n V_{\mathrm{sl}\,n} \oint_{S_1}(\mathbf{E}_{\mathrm{sl}\,n} \times \mathbf{H}_{pm}) \cdot \mathbf{n}ds . \qquad (2.51)$$

For the cavity $(p + 1)$, we have

$$\oint_{S_2}(\mathbf{E}_{p+1} \times \mathbf{H}_{p+1,m}) \cdot \mathbf{n}ds = \sum_n V_{\mathrm{sl}\,n} \oint_{S_2}(\mathbf{E}_{\mathrm{sl}\,n} \times \mathbf{H}_{p+1,m}) \cdot \mathbf{n}ds . \qquad (2.52)$$

Multiplying (2.49) by $\mathbf{H}_{\mathrm{sl}\,n}$ and integrating over the conditional boundary, we obtain

$$\left.\begin{aligned}
\oint_{S_1}(\mathbf{E}_{\mathrm{sl}} \times \mathbf{H}_{\mathrm{sl}\,n}) \cdot \mathbf{n}ds &= \sum_m V_{pm} \oint_{S_1}(\mathbf{E}_{pm} \times \mathbf{H}_{\mathrm{sl}\,m}) \cdot \mathbf{n}ds, \\
\oint_{S_2}(\mathbf{E}_{\mathrm{sl}} \times \mathbf{H}_{\mathrm{sl}\,n}) \cdot \mathbf{n}ds &= \sum_m V_{p+1,m} \oint_{S_2}(\mathbf{E}_{p+1,m} \times \mathbf{H}_{\mathrm{sl}\,m}) \cdot \mathbf{n}ds.
\end{aligned}\right\} \qquad (2.53)$$

Combining (2.45), (2.47), and (2.51)–(2.53), we obtain a set of equations in the unknown decomposition factors of the electric and magnetic fields in cavity p coupled with cavity $(p + 1)$ through a slot:

$$\left.\begin{aligned}
&\left[k_{pm}^2 - k^2\left(1 - \frac{j-1}{Q_{pm}}\right)\right]V_{pm} \\
&\quad + k_{pm}\sum_m V_{pm}C_p^{ml} + k_{pm}\sum_m V_{p+1,m}C_{p+1}^{ml} = 0, \\[2em]
&\left[k_{pm}^2 - k^2\left(1 - \frac{j-1}{Q_{pm}}\right)\right]I_{pm} \\
&\quad + j\omega\varepsilon_{\mathrm{a}}\sum_m C_p^{ml} + j\omega\varepsilon_{\mathrm{a}}\sum_m V_{p+1,m}C_{p+1}^{ml} = 0,
\end{aligned}\right\} \qquad (2.54)$$

where

$$
\left.
\begin{aligned}
C_p^{ml} &= \sum_l \frac{k_{\mathrm{sl}\,l} \displaystyle\int_{S_1} (\mathbf{E}_{\mathrm{sl}\,l} \times \mathbf{H}_{pm}) \cdot \mathbf{n}\, ds \displaystyle\int_{S_1} (\mathbf{E}_{pm} \times \mathbf{H}_{\mathrm{sl}\,l}) \cdot \mathbf{n}\, ds}{k_{\mathrm{sl}\,l}^2 - k^2\left(1 - \dfrac{j-1}{Q_{ml}}\right)}, \\[4ex]
C_{p+1}^{ml} &= \sum_l \frac{k_{\mathrm{sl}\,l} \displaystyle\oint_{S_1} (\mathbf{E}_{\mathrm{sl}\,l} \times \mathbf{H}_{pm}) \cdot \mathbf{n}\, ds \displaystyle\oint_{S_2} (\mathbf{E}_{p+1,m} \times \mathbf{H}_{\mathrm{sl}\,l}) \cdot \mathbf{n}\, ds}{k_{\mathrm{sl}\,l}^2 - k^2\left(1 - \dfrac{j-1}{Q_{ml}}\right)}.
\end{aligned}
\right\} \qquad (2.55)
$$

Here, the coefficients C_p^{ml} govern the frequency shift of the natural oscillation mode and the inter-mode coupling due to the effect of the coupling slot, and C_{p+1}^{ml} are the coefficients of inter-cavity coupling.

Thus, we obtained is $2M \times p$ system of algebraic equations, where M is the number of rotational eigenfunctions taken into account in the electromagnetic field expansion and p is the number of cavities in the system under consideration.

An analysis of (2.54) shows that all unknown decomposition factors depend on the sets of vector eigenfunctions for individual cells and on the cavities in the coupled system; they form a unified basis for expanding the total electromagnetic field.

One can solve set (2.54) for the unknown decomposition factors of electric the field V_{pm} and then use (2.36) to find the unknown decomposition factors of the magnetic field I_{pm} assuming that the relation between these factors remains the same as for free nondamped oscillations.

In the matrix form, we have

$$
\mathbf{A} \cdot \mathbf{V} = 0, \qquad (2.56)
$$

where $\mathbf{V}$ is the column-vector of the unknown decomposition factors and $\mathbf{A}$ is the matrix of the system with elements

$$A_{ij} = \begin{cases} k_i^2 - k^2\left(1 - \dfrac{j-1}{Q}\right) + k_iC_i^j, & i \neq j, \\[2mm] k_iC_i^j, & i = j, \end{cases} \tag{2.57}$$

where $i = 1, 2, ..., p$; $p = M \times N$; and M is the number of oscillation modes taken into account in the expansion of the field.

Thus, the determining of the resonance frequencies of a coupled system reduces to the eigenvalue problem for matrix $\mathbf{A}$. Because this matrix has complex-valued elements, the eigenvalues will also be complex. Their real part is related to the natural frequency of the coupled system, and the imaginary part is related to the attenuation in the system at this frequency.

The eigenvalues q_m of matrix $\mathbf{A}$ are determined as roots of the equation

$$\Delta(q) = 0, \tag{2.58}$$

where Δ is the matrix determinant. As follows from (2.54), this determinant is a polynomial of order $2 \times M \times N$ and, hence, has $2 \times M \times N$ complex conjugated roots.

We represent the determinant $\Delta(q)$ as an analytical function $G(q)$ having $2 \times M \times N$ roots. Since $G(q)$ is continuous, we may use the Newton–Raffson procedure [85] to compute these roots. Despite the fact that this procedure was designed for real-valued functions of a real argument, it is applicable (see [61]) to complex-valued functions of a complex argument. Following this procedure, we choose the first (initial) value $q^{(0)}$ of the desired root q and then improve it using the iteration formula

$$q^{(n+1)} = q^{(n)} - \frac{G(q^{(n)})}{G'(q^{(n)})}, \tag{2.59}$$

where n is the iteration number and $G'(q^{(n)})$ is the derivative of the complex-valued function G at point $q^{(n)}$.

An appropriate choice for the real part of the initial approximation is the natural frequency of one of coupled cavities, and for the imaginary part, the quantity inverse to the quality factor.

Iterations (2.59) continue until the difference between two successive approximations becomes smaller than a given value.

This procedure determins one of the roots of the determinant. In order to determine other roots, we must compensate for the found root in the function G. In so doing, we define the function G by the formula

$$G(q) = \frac{\Delta(q)}{\prod_{n=1}^{m}(q - q_n)}, \qquad (2.60)$$

which is appropriate for determining the $(m + 1)$th root.

Following this procedure, one can calculate all complex-valued roots. Their real and imaginary parts specify the natural frequencies and attenuation in the coupled system.

In order to compute the amplitude and phase distributions over the coupled system cavities, one must specify the non-zero right-hand side in the system (2.54):

$$\mathbf{A} \cdot \mathbf{V} = \mathbf{B}, \qquad (2.61)$$

where $\mathbf{B}$ is the $M \times N$ column-vector in which the only nonzero element corresponds to the cavity excited by an rf source.

Numerical solvers of systems of linear equations are classified as direct and iteration procedures. The elimination procedure is direct and has the advantage that it is finite and, theoretically, it can be applied for solving any non-singular set. Iteration procedures converge only for certain special systems.

To solve the system (2.61), we use a direct elimination procedure with the partial choice of the leading element [55]. For sets with poorly conditioned matrices, this procedure is as fast as the conventional Gauss method and more accurate than the Gauss method.

2.4   Equivalent Circuit Analysis

An alternative approach is to represent the designed accelerating structure by a chain of coupled circuits (CC) [86, 87]. This model is appropriate for describing the fields and computing the dynamic range of the accelerating structure. It needs only a few experimental electrodynamic parameters of individual cavities and coupling elements. An advantage of the CC model is that it does not require specific dimensions, so that conclusions drawn with this model are rather general. The theory is based on a mathematical formalism well developed in radio engineering for describing chains of coupled circuits. The load due to the beam can be included using the method of an equivalent generator of induced current or the concept of electron conductivity which is widely used in microwave electronics.

A limitation of the CC model is that it reduces the frequency band of the accelerating structure composed of many cells to the main passband. Furthermore, this model cannot be used for an integral rf system analysis. For example, with this model one cannot deduce the coupling properties from the geometry of the coupling slots and the matching iris aperture of the rf feeding waveguide.

Consider a chain of circuits that models an accelerating cavity composed of different slotted cells. Each cell can be considered as a series LCR circuit with inductance $2L_n$, capacitance C_n, and active resistance r_n. The intercircuit coupling is characterized by the coupling coefficient

$$k_{c\,n} = M \big/ \sqrt{L_n L_{n+1}} \;,$$

where L_n and L_{n+1} are the inductances of coupled coils (one half of the inductances of coupled circuits in this particular case) and M is the mutual inductance. The inductance of coupling elements is included into the inductance of the corresponding cell.

Figure 2.1 shows the equivalent circuit for a chain of cavities with coupling slots. The model takes into account only the coupling between adjacent cells. The pth cell includes the

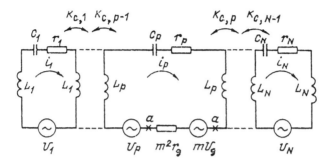

Figure 2.1. Equivalent circuit of a biperiodic structure (BPS)

coupling with the waveguide that feeds rf power from a generator. The feeding waveguide is simulated as an ideal transformer with transformation ratio m. The inductance and resistance of the coupling element is included into the corresponding parameters of the cell. The rf-power generator matched with the transmitting line is shown as an equivalent voltage generator U_g with internal resistance r_g. Voltage generators U_p take into account the effect of accelerated beam [88].

This equivalent circuit is valid within the frequency range where the above parameters remain invariable. Higher oscillation modes in cells and resonance frequencies of coupling elements must not fall into this frequency range [12, 86]. Obviously, this circuit is applicable in the frequency range with TM_{010} mode in the cells. However, it was successfully used for the TM_{020} mode [89].

Writing Kirchhoff's equations for every loop of the circuit shown in Fig. 2.1, we obtain a set of N equations for N unknown currents:

$$\left(2j\,\omega L_1 + r_1 + \frac{1}{j\omega C_1} \right) i_1 + k_{c\,1} j\omega \sqrt{L_1 L_2}\, i_2 = U_1,$$

....

$$\left(2j\omega L_n + r_n + \frac{1}{j\omega C_n}\right) i_n + k_{c\,n\text{-}1} j\omega\sqrt{L_{n-1}L_n}\; i_{n-1}$$

$$+ k_{c\,n} j\omega\sqrt{L_n L_{n+1}}\; i_{n+1} = U_n,$$

....

$$\left(2j\omega L_p + r_p + \frac{1}{j\omega C_p}\right) i_p + k_{c\,p\text{-}1} j\omega\sqrt{L_{p-1}L_p}\; i_{p-1} \qquad\qquad (2.62)$$

$$+ k_{c\,p} j\omega\sqrt{L_p L_{p+1}}\; i_{p+1} = U_p + U_{aa},$$

....

$$\left(2j\omega L_N + r_N + \frac{1}{j\omega C_N}\right) i_N + k_{c\,N\text{-}1} j\omega\sqrt{L_N L_{N-1}}\; i_{N-1} = U_N.$$

Here, U_{aa} is the complex voltage at terminals a, a.

We rewrite the set (2.62) in the form

$$X_1\Phi_1 + k_{c\,1} X_2/2 = J_1,$$

....

$$X_n\Phi_n + k_{c\,n-1} X_{n-1}/2 + k_{c\,n+1} X_{n+1}/2 = J_n,$$

....

$$X_p\Phi_p + k_{c\,p-1} X_{p-1}/2 + k_{c\,p+1} X_{p+1}/2 = J_p + \frac{U_{aa}}{i\omega\sqrt{2L_p}}, \qquad (2.63)$$

....

$$X_N\Phi_N + k_{c\,N\text{-}1} X_{N-1}/2 = J_N,$$

where

$$\Phi_n = 1 - \left(\frac{\omega_n}{\omega}\right)^2 + \frac{1}{jQ_n}\frac{\omega_n}{\omega},$$

$$X_n = i_n\sqrt{2L_n},$$

$$J_n = U_n \Big/ \left(j\omega\sqrt{2L_n}\right),$$

ω_n and Q_n are the resonance frequency and unloaded quality factor of the nth loop, and $(1/2)|X_n|^2$ is the energy stored in the nth loop.

Set (2.63) depends now on parameters ω_n and Q_n that can be determined experimentally.

We rewrite the second term on the right-hand side of the pth equation in the form

$$\frac{U_{aa}}{j\omega\sqrt{2L_p}} = \frac{Z_{aa}}{j\omega 2L_p}X_p = \frac{1}{jQ_p}\frac{\beta_p}{\beta}X_p,\qquad(2.64)$$

where $Z_{aa} = U_{aa}/i_p$ is the complex impedance of the equivalent circuit across terminals a–a; $\beta = m^2 r_g/Z_{aa}$ is the total normalized conductance of the equivalent circuit at points a, a (this quantity is real at the resonance frequency); and $\beta_p = m^2 r_g/r_p$ is the coefficient of coupling of an isolated cell with the feeding waveguide.

The total power across Z_{aa} can be presented as

$$S_{aa} = \frac{U_{aa}i_p^*}{2} = \frac{4m^2 r_g P_g}{|m^2 r_g + Z_{aa}|^2}Z_{aa},$$

so that

$$\frac{8P_g}{\omega|1+1/\beta|^2} = |X_p|^2\frac{\beta_p}{Q_p}.\qquad(2.65)$$

For given β_p and Q_p, this formula relates the generator power P_g to the field in the pth cell and quantity β.

Thus, expressions (2.63)–(2.65) form a set that describes the fields in the CC and the normalized conductance at terminals a, a (which is the impedance function of an accelerating cavity) in terms of the given rf power P_g, coupling coefficient β_p, and quantities J_n. The latter quantities are representable as [88]

$$J_n = \frac{I_{in}}{j\omega} \sqrt{\frac{r_n \omega_n}{Q_n}} , \qquad (2.66)$$

where I_{in} is the accelerated first harmonic of the induced current.

In the general case, J_n is a nonlinear function depending on the motion of particles in the accelerating field of the cell. Therefore, the system (2.63) with the right-hand side (2.66) is nonlinear, and can be resolved by numerical methods.

In some particular cases, this system can be solved analytically. As an example, we consider a biperiodic accelerating structure excited in the $\pi/2$ mode. The phase velocity of the wave coincides with the velocity of the bunch and is given by the formula $\beta_{ph} = 2(L_a + L_c)/\lambda$, where L_a and L_c are the lengths of accelerating and coupling cells, respectively. The quality factors of accelerating cells (cells with odd numbers) are identical $(Q_{2n+1} = Q_a)$ as well as the quality-factors of coupling cells $(Q_{2n} = Q_c)$. The frequencies of accelerating and coupling cells (coupling apertures included) also coincide $\omega_{2n+1} = \omega_{2n} = \omega$. Along the structure, the coupling coefficient k_c remains invariable, and $r_1 = r_3, \ldots, r_N = r_a$. The accelerating beam is a series of bunches with a constant phase distribution that follow each other with a period corresponding to the rf frequency. Because phase motions are absent in such a beam, the first harmonic of the induced current will be identical for all accelerating cells. We will neglect the excitation of coupling cells because the fields in these cells are small in comparison to the fields in accelerating cells. Under these conditions, J_n can be represented as [88]

$$J_n = \begin{cases} -\dfrac{X_n}{jQ_a} A, & n = 1,3,5,\ldots,N; \\[2mm] 0, & n = 2,4,6,\ldots N-1, \end{cases} \qquad (2.67)$$

where

$$A = \frac{\alpha e^{j\varphi_i}}{(1-\alpha)\cos\varphi_i} ,$$

$$\alpha = \frac{P_I}{P_I + P_A} = \frac{-r_a|I_i|\cos\varphi_i}{X_p\sqrt{r_a\,\omega/Q_a} - r_a|I_i|\cos\varphi_i},$$

P_I is the increment of the beam power in the pth cell, P_A is the power lost in the walls, and φ_1 is the phase difference between the first harmonics of the induced current and the complex amplitude of the voltage across the circuit.

Under the above assumptions, (2.63) takes the form

$$
\left.
\begin{aligned}
&\frac{1}{jQ_a}X_1 + \frac{k_c}{2}X_2 = -\frac{A}{jQ_a}X_1,\\[2mm]
&\frac{1}{jQ_c}X_2 + \frac{k_c}{2}X_1 + \frac{k_c}{2}X_3 = 0,\\[2mm]
&\quad\cdots\cdots\cdots\cdots\cdots\cdots\cdots\cdots\cdots\\[1mm]
&\frac{1}{jQ_a}X_{2n-1} + \frac{k_c}{2}X_{2n-2} + \frac{k_c}{2}X_{2n} = -\frac{A}{jQ_c}X_{2n-1},\\[2mm]
&\frac{1}{jQ_c}X_{2n} + \frac{k_c}{2}X_{2n-1} + \frac{k_c}{2}X_{2n+1} = 0,\\[2mm]
&\frac{1}{jQ_a}X_p + \frac{k_c}{2}X_{p-1} + \frac{k_c}{2}X_{p+1} = -\frac{A}{jQ_a}X_p + \frac{\beta_p}{jQ_a\beta}X_p,\\[2mm]
&\quad\cdots\cdots\cdots\cdots\cdots\cdots\cdots\cdots\cdots\\[1mm]
&\frac{1}{jQ_a}X_N + \frac{k_c}{2}X_{N-1} = -\frac{A}{jQ_a}X_N.
\end{aligned}
\right\}
\qquad (2.68)
$$

An approximate solution to this system be written as

$$
\left.
\begin{aligned}
&X_{2n-1} \approx (-1)^{n+1}\left[1 + \frac{2(n^2-n)}{k_c^2\,Q_a Q_c}(1+A)\right]X_1,\\[3mm]
&X_{2n} \approx (-1)^{n+1}\frac{j\,2n}{k_c Q_a}(1+A)X_1,
\end{aligned}
\right\}
\qquad (2.69)
$$

where $n = 1, 2, 3, ..., (N-1)/2$.

We included only terms of the order of $(k_c^2 Q_a Q_c)^{-1}$ in the first row of expression (2.69) and of the order of $(k_c Q_a)^{-1}$ in the second row. This inclusion is justified because $(k_c^2 Q_a Q_c)^{-1} \ll 1$. Moreover, the inequality

$$\left| \frac{2(n^2 - n)(1 + A)}{k_c^2 Q_a Q_c} \right| \ll 1$$

holds for short cells and of low beam loading.

If the beam current vanishes ($A = 0$), then

$$
\left.
\begin{aligned}
X_{2n-1} &\approx (-1)^{n+1} \left[1 + \frac{2n(n-1)}{k_c^2 Q_a Q_c} \right] X_1, \\[2mm]
X_{2n} &\approx (-1)^{n+1} j \frac{2n}{k_c Q_a} X_1.
\end{aligned}
\right\}
\tag{2.70}
$$

For a cavity with half-cells at the ends, the second term on the left-hand side of the first equation in (2.68) acquires an additional factor of 2. In the absence of the beam loading, the corresponding solutions will be

$$
\left.
\begin{aligned}
X_{2n-1} &\approx (-1)^{n+1} \left[1 + \frac{2(n-1)^2}{k_c^2 Q_a Q_c} \right] X_1, \\[2mm]
X_{2n} &\approx (-1)^{n+1} j \frac{2n-1}{k_c Q_a} X_1.
\end{aligned}
\right\}
\tag{2.71}
$$

An analysis of expressions (2.70) and (2.71) shows that, in short structures, the field in accelerating cells can be deemed independent of the cell number (or the point, where the rf power is fed). In coupling cells, the field linearly increases to the point, where rf power is fed, but its maximum is smaller by a factor of $k_c Q_a$. Expressions (2.69) and (2.70) show that, in biperiodic structures, the accelerating field in the $\pi/2$ mode is stable to disturbances because, in the first approximation, losses and beam loading do not shift the phases of the fields in the accelerating cells.

In order to obtain the magnitude of the field in the cells, one must solve the system (2.68) together with equation (2.65). Normalizing (2.68) by X_p ($X_p \neq 0$), we obtain two independent systems for cell p

$$\left.\begin{array}{l} \dfrac{1}{jQ_a}x_1 + \dfrac{k_c}{2}x_2 = -\dfrac{A}{jQ_a}x_1, \\[4mm] \cdots\cdots\cdots\cdots\cdots\cdots\cdots\cdots\cdots\cdots\cdots \\[4mm] \dfrac{1}{jQ_a}x_{2n-1} + \dfrac{k_c}{2}x_{2n-2} + \dfrac{k_c}{2}x_{2n} = -\dfrac{A}{jQ_a}x_{2n-1}, \\[4mm] \dfrac{1}{jQ_c}x_{2n} + \dfrac{k_c}{2}x_{2n-1} + \dfrac{k_c}{2}x_{2n+1} = 0, \end{array}\right\} \quad (2.72a)$$

$$\left.\begin{array}{l} \dfrac{1}{jQ_c}x_{p-1} + \dfrac{k_c}{2}x_{p-2} = -\dfrac{k_c}{2}, \\[4mm] \dfrac{1}{jQ_a} + \dfrac{k_c}{2}x_{p-1} + \dfrac{k_c}{2}x_{p+1} = -\dfrac{A}{jQ_a} + \dfrac{\beta_p}{jQ_a\beta}, \\[4mm] \dfrac{1}{jQ_c}x_{p+1} + \dfrac{k_c}{2}x_{2p+2} = -\dfrac{k_c}{2}, \end{array}\right\} \quad (2.72b)$$

$$\left.\begin{array}{l} \dfrac{1}{jQ_a}x_{2n-1} + \dfrac{k_c}{2}x_{2n-2} + \dfrac{k_c}{2}x_{2n} = -\dfrac{A}{jQ_a}x_{2n-1}, \\[4mm] \dfrac{1}{jQ_c}x_{2n} + \dfrac{k_c}{2}x_{2n-1} + \dfrac{k_c}{2}x_{2n+1} = 0, \\[4mm] \cdots\cdots\cdots\cdots\cdots\cdots\cdots\cdots\cdots\cdots\cdots \end{array}\right\} \quad (2.72c)$$

$$\frac{1}{jQ_a} x_N + \frac{k_c}{2} x_{N-1} = -\frac{A}{jQ_a} x_N,$$

where $x_n = X_n/X_p$. In (2.72a), n runs from 1 to $(p-3)/2$ and, in set (2.72c), n runs from $(p+3)/2$ to $(N+1)/2$. The quantities x_{p-1} and x_{p+1} are the solutions of two independent sets (2.72a) and (2.72c).

To determine x_{p-1}, we resort to the general solution (2.69). For set (2.72a), these solutions have the form

$$x_{p-1} = (-1)^{\frac{p+1}{2}} j \frac{(p-1)}{k_c Q_a}(1+A)x_1, \quad x_{p-2} = (-1)^{\frac{p+1}{2}} x_1.$$

Substituting these solutions into the last equation of (2.72a), we obtain

$$x_{p-1} = -j\frac{(p-1)}{k_c Q_a}(1+A) . \qquad (2.73)$$

Similarly, we find from (2.72c)

$$x_{p+1} = -j\frac{(N-p)}{k_c Q_a}(1+A) . \qquad (2.74)$$

Substituting (2.73) and (2.74) into (2.72b), we obtain

$$\beta = \frac{2\beta_p}{(N+1)(1+A)} . \qquad (2.75)$$

Using the expression for α, we find

$$\beta = \frac{\beta_p(1-\alpha)}{N_a(1+j\alpha\tan\varphi_i)} , \qquad (2.76)$$

where $N_a = (N+1)/2$ is the number of accelerating cells in the cavity. If $\alpha = 0$, then $\beta = \beta_0 = \beta_p/N_a$ is the real-valued quantity characterizing the initial coupling between the cavity and waveguide. Now, we can write

$$\beta = \beta_0 \frac{1-\alpha}{(1+j\alpha \tan \varphi_i)} . \tag{2.77}$$

Using (2.77), we rearrange expression (2.65) to obtain the magnitude of the field in the cell p:

$$|X_p| = \frac{2(1-\alpha)}{|\beta_0(1-\alpha)+1+j\alpha \tan \varphi_i|} \sqrt{\frac{2P_g \beta_0 Q_p}{\omega N_a}} . \tag{2.78}$$

The fields in other cells are determined as

$$X_n = |X_p| x_n .$$

Thus, with the CC model, one can determine the fields in regular biperiodic structures taking into account such factors as losses, effects of beam loading (in the linear approximation), and the impedance function of the accelerating cavity with the allowance for the beam current. It is worth noting that the CC model can be used also to analyze the stability of the accelerating field distribution [89].

3

COUPLERS AND COUPLING ELEMENTS

3.1 General

Coupling slots between adjacent cavities can break the axial symmetry of the electromagnetic fields in these structures. An on-axis coupled structure (biperiodic structure or BPS) is an illustrative example (see Fig. 1.4b). Asymmetry also occurs in the couplers connecting these structures with a feeding waveguide. Rigorous calculations of electromagnetic fields in such accelerating structures require three-dimensional numerical techniques.

The coupling elements between adjacent cells of biperiodic structures are usually made as narrow azimuthal slots cut in the irises. In the equations (2.38), the cavity coupling is described by the integral over the coupling element surface

$$\int_{S_s} (\mathbf{E}_p \times \mathbf{H}_{mp}) \cdot \mathbf{n} \, ds \,. \tag{3.1}$$

It is obvious that the modeling of a coupling element reduces to choosing a method of obtaining a specific expression for (3.1).

An efficient method of modeling narrow azimuthal slots of BPSs lies in their representation in the form of a lossless strip line that is short-circuited at the ends [90, 91]. In this chapter, we will consider this model as well as a simpler model based on representing coupling slots in the form of a radiating antenna.

To design a coupler between accelerating structures and a feeding rf section, one can resort to available experimental reference data or use approximate analytic methods. Reference data are available for a coupler that transforms waves from a rectangular waveguide to a DLW [3, 4]. For resonant cavities, including BPSs, it is useful to divide a complicated electrodynamic system into two domains: a cavity system, whose electrodynamic model was discussed in Chapter 2, and a feeding waveguide section. Using the continuity condition for the electric or magnetic field at the boundary yields a set of equations describing the electrodynamic system as a whole [48]. In what follows, this method will be applied to an rf power coupler of a BPS. We also present a technique to calculate waveguide feeders to accelerating cavities—the cavity-analogue method.

We calculate the electric field at the center of an accelerating gap of an Ω-shaped cavity to illustrate the possibilities of the cavity-analogue method. Figure 3.1 shows the calculated distribution of the magnetic field. The length of the cylindrical cavity is equal to that of the Ω-shaped cavity and the frequencies of both resonators are identical ($f_0 = 2475$ MHz). An analysis of these distributions and a rapid decay of the H_φ-component of the field with the distance from the wall suggests that the magnetic flux through the cross section of the Ω-shaped cavity, which determines the voltage U at the accelerating gap, may be replaced by the corresponding magnetic flux for the cavity-analogue. Then

$$U(r=0) = \mu_0 \omega_0 \int_0^L \int_0^R H_\varphi(r,z)\, dr\, dz \, ,$$

and we finally obtain

Figure 3.1. Magnetic field in the Ω-shaped cavity (*1*) at the end surface and (*2*) in the middle plane and (*3*) in the cylindrical cavity excited in the TM$_{010}$ mode.

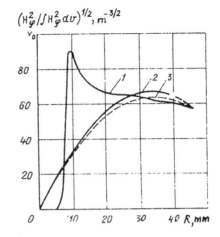

$$\frac{U}{\sqrt{PQ}} = \frac{2}{v_{01}J_1(v_{01})} \sqrt{\frac{Z_0 L}{\lambda_0}} = 1.602 \sqrt{\frac{Z_0 L}{\lambda_0}} . \qquad (3.2)$$

The axial distribution of the electric field in the Ω-shaped cavity is approximated as $E_z(0, z) = E_0$ within the gap $-d_g/2 \le z \le d_g/2$ and as $E_z(0, z) = E_0 \exp[-v_{01}(z - d_g/2)/a]$ for $z \ge |d_g/2|$, i.e., as in a cutoff waveguide of radius a. Equating the voltage at the accelerating gap of the Ω-shaped cavity

$$U = E_0 d_g + 2 \int_{d_g/2}^{\infty} E_0 \exp\left[-v_{01}\left(z - d_g/2\right)/a\right] dz$$

to that at the cavity-analogue, we obtain

$$\frac{E_0}{\sqrt{PQ}} = \frac{1.602}{d_g + 2a/v_{01}} \sqrt{\frac{Z_0 L}{\lambda_0}} . \qquad (3.3)$$

Substituting $L = 4.9 \cdot 10^{-2}$ m, $d_g = 2.98 \cdot 10^{-2}$ m, $a = 5 \cdot 10^{-3}$ m, $\lambda_0 = 0.121$ m, and $v_{01} = 2.405$ from [92], we find $E_0/\sqrt{PQ} = 582$ $\Omega^{1/2}$/m.

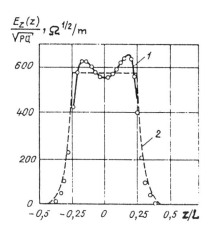

Figure 3.2. Measured (*1*) and calculated (*2*) axial distributions of the electric-field parameter in Ω-shaped cavity.

Figure 3.2 shows the measured and calculated axial distributions of the electric field parameter in the Ω-shaped cavity. The fact that at the center of the gap this parameter is predicted with an accuracy of 2% or better justifies the use of the cavity-analogue method for solving a number of problems, including couplers and coupling slots.

3.2 Coupling Slots

First, we will consider coupling slots between BPS cells by representing these narrow azimuthal slots in the form of a lossless strip line short-circuited at its ends. Within the framework of this model, we may neglect the curvature of the coupling slot and switch to a Cartesian coordinate system with its origin at the center of the slot with dimensions l_s and Δ along the x- and y-axes, respectively. In this case, the voltage across the slot satisfies the equation

$$\left(\frac{\partial^2}{\partial x^2} + k^2\right) U_s(x) = -F(x), \tag{3.4}$$

where $F(x)$ is a forcing function.

In a single-mode approximation, the magnetic field is uniform over the surface of narrow azimuthal slots (azimuthal variations not included) and

$$F(x) = j\omega L_0\left(I_p H_p - I_{p\pm1}H_{p\pm1}\right), \tag{3.5}$$

where L_0 is the inductance of the strip line per unit length. Physically, (3.5) means that the slots are excited by the difference of the magnetic fields of adjacent cavities.

The solution of the differential equation (3.4) can be written in the form

$$U_s(x) = j\omega L_0\left[\frac{\cos kx}{\cos(kl_s/2)} - 1\right]\left(I_p H_p - I_{p\pm1}H_{p\pm1}\right),$$

so that

$$\int_{S_s}(\mathbf{E}_p \times \mathbf{H}_p)\cdot\mathbf{n}\,ds = H_p\int_{-l_s/2}^{l_s/2}\int_{-\Delta/2}^{\Delta/2}E_\tau\,dx\,dy = H_p\int_{-l_s/2}^{l_s/2}U_s(x)\,dx$$

$$= -j\frac{\omega L_0 l_s}{k^2}\left[\frac{\tan(kl_s/2)}{kl_s/2} - 1\right]\left(I_p H_p^2 - I_{p\pm1}H_{p\pm1}H_p\right).$$

Substituting this expression into equation (2.45) and letting $Q_p \to \infty$, we obtain for cell p:

$$\left(k_p^2 - k^2\right)I_p - c^2 L_0\varepsilon_0 l_s\left[\frac{\tan(kl_s/2)}{kl_s/2} - 1\right]$$

$$\times\left(2H_p^2 I_p - H_p H_{p+1}I_{p+1} - H_p H_{p-1}I_{p-1}\right). \tag{3.6}$$

The BPS cells with the numbers $p-2$ and $p+2$ are identical; i.e., $H_{p-1} = H_{p+1}$, $H_p = H_{p+2}$, etc. Using the notation

$$\Phi = c^2 L_0\varepsilon_0 l_s\left[\frac{\tan(kl_s/2)}{kl_s/2} - 1\right]$$

we can rewrite (3.6) as

$$\left(1 - \frac{k_p^2}{k^2} + \frac{2\Phi}{k^2}\right)I_p = \frac{\Phi H_{p+1} H_p}{k^2}\left(I_{p+1} + I_p\right). \qquad (3.7)$$

Relation (3.7) represents the dispersion equation of a BPS and it should be compared to a similar equation derived for a chain of coupled circuits [86, 88]

$$\left(1 - \frac{k_{0p}^2}{k^2}\right)X_p + \frac{k_s}{2}\left(X_{p-1} + X_{p+1}\right) = 0,$$

where k_{0p} is the wavenumber of the pth circuit and $1/2(X_p)^2$ is the energy stored in the circuit. This comparison yields

$$k_c = \frac{2c^2 L_0 \varepsilon_0 l_s}{k^2}\left[1 - \frac{\tan(kl_s/2)}{kl_s/2}\right]H_p H_{p+1}, \qquad (3.8)$$

$$k_p^2 = k_{0p}^2 + 2c^2 L_0 \varepsilon_0 l_s\left[1 - \frac{\tan(kl_s/2)}{kl_s/2}\right]H_p^2, \qquad (3.9)$$

where k_p is the wavenumber of the pth cavity with a perturbation from coupling slots taken into account; k_{0p} is the wavenumber of the same cavity without coupling slots; and $k_{p-1} = k_p = k_{p+1} = \dots = 2\pi/\lambda_p$ for $\theta = \pi/2$ mode in the case of a symmetric dispersion curve. Using the relations $\varepsilon_0 c = Z_0^{-1}$ and $Z_s = L_0 c$, where Z_s is the wave impedance of the strip line, extended to the case of N coupling slots of adjacent cells, it is appropriate to rewrite formulas (3.8) and (3.9) in the form [92]

$$\left.\begin{array}{l} k_c = \dfrac{2Z_s l_s}{Z_0 k_p^2}\left[1 - \dfrac{\tan(kl_s/2)}{kl_s/2}\right]N H_p H_{p+1}, \\[4mm] f_p = f_{0p}\left(1 - k_c \dfrac{H_p}{H_{p+1}}\right)^{-1/2}. \end{array}\right\} \qquad (3.10)$$

Denoting the thickness of the wall between BPS cells by t (Fig. 3.3), we obtain the wave impedance of a symmetric strip line in the range $0.6 \leq \Delta/t \leq 3$ from the relation

$$\frac{Z_s}{Z_0} = \left\{ \frac{t}{\Delta} + \frac{2}{\pi} \left[1 + \ln\left(\frac{\pi t}{2\Delta} + 1 \right) \right] \right\}^{-1}. \qquad (3.11)$$

To find H_p and H_{p+1}, it is necessary to invoke both the orthonormalization condition

$$H_p = \sqrt{\frac{H_p^2(r_s)\mu_0}{2W}}, \qquad (3.12)$$

where $H_p^2(r_s)$ is the magnetic field of cavity p at the slot coordinate r_s, and the expressions for the fields of a cavity-analogue, which are identical for $r_s/R > 0.6$ to the field of the Ω-shaped accelerating cavity:

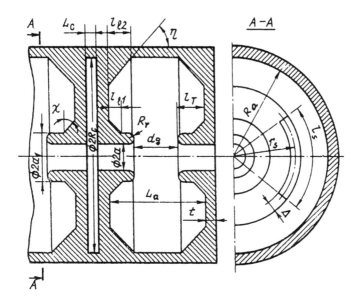

Figure 3.3. On-axis coupled structure with coupling slots.

$$H_p \cong \frac{J_1(\nu_{01} r_s/R) k_p}{\pi J_1(\nu_{01}) \sqrt{\pi L_p}},$$

$$H_p H_{p+1} \cong \frac{J_1^2(\nu_{01} r_s/R) k_p^2}{\pi J_1^2(\nu_{01}) \nu_{01}^2 \sqrt{L_p L_{p+1}}}. \quad \left. \right\} \tag{3.13}$$

The index p in (3.13) can denote both an accelerating cell (subscript a) and a coupling cell (c). Accurate to the first two terms of the expansion $\tan x = x + x^3/3 + \dots$, relations (3.10) have the form

$$k_c = -\frac{Z_s}{Z_0} N \frac{l_s^3}{6} H_c(r_s) H_a(r_s),$$

$$f_{a,c} = f_0 \left[1 + H_{a,c}^2(r_s) \frac{l_s^3}{6} N \frac{Z_s}{Z_0} \right]^{-1/2}. \quad \left. \right\} \tag{3.14}$$

It should be remembered that $L_a = D - L_c - 2t$ and D is the BPS period.

Relations (3.11), (3.13), and (3.14) form the basis for calculating the size of coupling slots from a specified coupling coefficient and for considering their effect on the frequencies of the cells. Figure 3.4 shows the corresponding plots. In practice, face milling is used to cut coupling slots, so that the ends of the slots are rounded off with radius $R_{cs} = \Delta/2$. For this reason, calculations must be performed for a reduced slot length

$$l_s = l_{cs}(1 - 0.236 \, \Delta/l_{cs}). \tag{3.15}$$

Similar results relating to the effect of coupling slots can also be obtained in the context of a simpler model representing a coupling slot in the form of an antenna. Accordingly, a slot that is cut in a conducting screen confining a uniform magnetic field is equivalent to a magnetic dipole. In this case [93], we have

$$I^m l^m = j \frac{2\pi Z_0}{\lambda_0} (H_{r1} - H_{r2}) p_m e^{-\alpha t}, \tag{3.16}$$

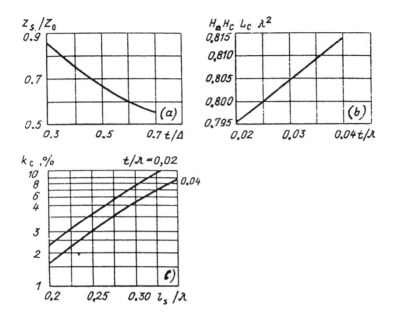

Figure 3.4. (a) Wave impedance of a coupling slot as a function of t/Δ; (b) the plot corresponding to relation (3.13) at $\Delta/\lambda = 0.069$, $L_a/L_c = 9.5$ and $r_s/R_a = 0.65$; and (c) the coupling coefficient of an optimized BPS as a function of l_s/λ for different values of the parameter t/λ, and at $L_c/\lambda = 0.04$ and $\beta_{ph} = 1$.

where I^m is an magnetic current; l^m is the dipole length; $H_{\tau 1}$ and $H_{\tau 2}$ are the tangential components of the magnetic field to the right and to the left from the slot, respectively; t is the thickness of the screen; and p_m is the magnetic polarizability of the slot. In accordance with the principle of equivalence of the sources

$$I^m l^m = j^m S_s = -\mathbf{n} \cdot \mathbf{E} S_s, \qquad (3.17)$$

where j^m is the surface density of the magnetic current and S_s is the slot area. Thus, the surface integral (3.1) is calculated in the form

$$\int_{S_s} (\mathbf{E}_p \times \mathbf{H}_{mp}) \cdot \mathbf{n} \, ds = -jk_p Z_0 p_M \left(H_p I_p - I_{p+1} H_{p+1} - I_{p-1} H_{p-1} \right) N e^{-\alpha t}.$$

$$(3.18)$$

Rearranging (3.18) as above yields expressions for the coupling coefficient. We also substitute the magnetic polarizability of a narrow slot

$$p_{\mathrm{m}} = \pi l_s^3 \left\{ 24 \left[\ln(4 l_s / \Delta) - 1 \right] \right\}^{-1}$$

borrowed from [93] and the attenuation coefficient

$$\alpha = 2 \pi / \lambda_{\mathrm{cr}} = k_p \sqrt{\lambda_p^2 / (2 l_s)^2 - 1}$$

corresponding to a cutoff rectangular waveguide with a TE_{10} mode. Finally, for the antenna model of a coupling slot, we obtain

$$k_{\mathrm{c}}^{(A)} \cong \frac{\pi^2 J_1^2(v_{01} r_s / R)}{3 v_{01}^2 J_1^2(v_{01}) \lambda_p^2} \frac{l_s^3 N}{\sqrt{L_y L_{\mathrm{c}}}} \left(\ln \frac{4 l_s}{\Delta} - 1 \right)^{-1} \exp \left(-\frac{2 \pi t}{\lambda_p} \sqrt{\frac{\lambda_p^2}{4 l_s^2} - 1} \right).$$

$$(3.19)$$

The ratio of the coupling coefficients calculated by formulas (3.19) and (3.14) is

$$\frac{k_{\mathrm{c}}^{(A)}}{k_{\mathrm{c}}} = \frac{1 + \pi t / (2\Delta) + \ln \left[1 + \pi t / (2\Delta) \right]}{\ln(4 l_s / \Delta) - 1} \exp \left(-\frac{2 \pi t}{\lambda_p} \sqrt{\frac{\lambda_p^2}{4 l_s^2} - 1} \right). \quad (3.20)$$

Formula (3.20) is tabulated in Table 3.1, where it is assumed that $\Delta/\lambda_p = 0.06$. This table demonstrates that the models of coupling slots considered above are equivalent within the range $0.03 \le t/\lambda_p = 0.04$, which is typical for practical applications. The dipole model algorithm (3.19) has a wider area of application than the strip model, because the magnetic polarizability p_{m} is known for a wide variety of coupling slot geometries such as circles, ovals, crosses, etc. One more advantage of this model is associated with its applicability to structures with electrically coupled cells.

Table 3.1. A comparison of two models of the coupling coefficient for different values of l_s/λ_p.

t/λ_p	l_s/λ_p					
	0.15	0.20	0.25	0.30	0.35	0.40
0.03	0.997	0.966	0.941	0.923	0.908	0.900
0.04	0.956	0.977	0.987	0.991	0.995	1.002

Calculations of $k_c = F(l_s/\lambda_p, t/\lambda_p, \Delta/\lambda_p, L_c/\lambda_p, L_a/\lambda_p)$ and $f_c/f_{0c} = F(l_s/\lambda_c, l_c/\lambda_c, t/\Delta)$ were verified with experimental mockups. Figure 3.5 shows the results for typical calculations.

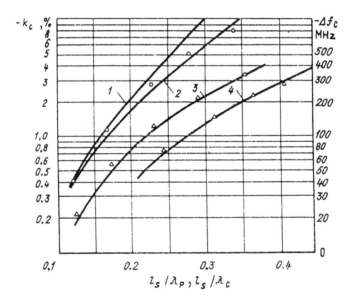

Figure 3.5 Coupling coefficient (curves *1* and *2*) and the frequency deviations of coupling cells (curves *3* and *4*) as functions of coupling slot size at $t/\Delta = 0.667$ and $r_s/R = 0.68$. Measured data are shown by circles for k_c and by triangles for Δf_c.

Curves *1* and *2* for the coupling coefficient are constructed for the accelerating mockup of the Research Institute of Nuclear Physics at the Moscow State University [28]. Here, $t = 4$ mm, $L_a = 49$ mm, $f_{0a} = 2473$ MHz, $t/\Delta = 0.667$, and $r_s = 30$ mm. Curve *1* corresponds to formula (3.10) and curve *2* is obtained using the simplified relation (3.14). The experimental data confirm the simpler algorithm, especially for $l_s/\lambda_p > 0.2$. The reason for this is as follows. Formula (3.10) disregards the losses in a coupling slot near its resonance. In particular, this formula yields $k_c \rightarrow \infty$ as $l_s/\lambda_p \rightarrow 1/2$, which is not the case in practice. A method of calculating the coupling coefficient with allowance for a finite Q-factor is developed in [87]. An analysis of the results obtained in that work shows that the losses in a coupling slot are almost exactly compensated by the terms of order higher than three in the expansion

$$\tan\frac{k_p l_s}{2} = \frac{k_p l_s}{2} + \frac{1}{3}\left(\frac{k_p l_s}{2}\right)^3 + \ldots.$$

Consequently, formula (3.14) must be used in practical calculations. Dependence *3* (Fig. 3.5) of the frequency deviation of cylindrical coupling cells ($f_{0c} = 2530$ MHz) was plotted for the same structure. The calculated curve was worked out using formula (3.14). The dependence $\Delta f_c = f_c - f_{c0} = F(l_s/\lambda_c)$ (curve *4* in Fig. 3.5) spans the range of higher values of l_s/λ_c and approaches $l_s/\lambda_c = 0.5$. The calculated and measured data of the latter dependence correspond to $L_c = 4$ mm, $R_c = 38.5$ mm, and $f_{0c} = 2989$ MHz. The measurements were carried out with a single coupling slot, which corresponds to $N = 1/2$.

The results of testing do not cast any doubt on the validity of the analytic relations (3.14) and (3.19) for calculating the geometric dimensions of coupling slots at the stage of engineering design of BPSs.

3.3   The Effect of External Circuits

Consider an rf coupler that feeds rf power into a biperiodic structure (see Fig. 3.6). Passing a boundary through the surface of the iris connecting the feeding waveguide with the accelerating structure, we assume the waveguide and the accelerating structure to be the first and second partial domains, respectively. The thickness of the iris separating the waveguide and the structure is assumed to be negligible.

Suppose that the tangent electric field strength $\mathbf{E}_\tau$ on the surface of the matching iris aperture S_s is known. Then the electromagnetic field in the partial domains I and II can be determined from this electric field.

For the continuity of the field in the system, it is necessary that the tangent components of the magnetic field be continuous at the boundary; i.e., the surface density of the electric current be zero [48]:

$$\mathbf{H}^I(\mathbf{j}^I,\mathbf{E}_\tau)\times\mathbf{n}_{12}+\mathbf{H}^{II}(\mathbf{j}^{II},\mathbf{E}_\tau)\times\mathbf{n}_{21}=0\,, \qquad (3.21)$$

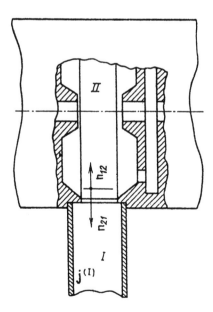

Figure 3.6. Biperiodic structure with a feeding waveguide.

Here $\mathbf{H}^I$ and $\mathbf{H}^{II}$ are the magnetic field strengths excited at the coupling iris aperture on the sides of the first and second domains by extraneous sources $\mathbf{j}^I$ and $\mathbf{j}^{II}$ and by the field $\mathbf{E}_r$, and $\mathbf{n}_{12}$ and $\mathbf{n}_{21}$ are the outer normals to the first and the second domain, respectively.

In equation (3.21), a microwave generator stands for the extraneous source $\mathbf{j}^I$. Assuming a decoupling between the generator and the cavity, we may write, by the superposition

$$\mathbf{H}^I(\mathbf{j}^I, \mathbf{E}_r) = \mathbf{H}^I(0, \mathbf{E}_r) + \mathbf{H}^I(\mathbf{j}^I, 0), \qquad (3.22)$$

where $\mathbf{H}^I(0, \mathbf{E}_r)$ is the magnetic field excited by $\mathbf{E}_r$ alone and $\mathbf{H}^I(\mathbf{j}^I, 0)$ is the magnetic field excited by $\mathbf{j}^I$ provided that the coupling iris surface is conducting ($\mathbf{E}_r = 0$). Taking into account (3.22) and simplifying the notation for the magnetic field strength in the cavity system, we represent equation (3.21) in the form

$$\mathbf{n}_{21} \times \mathbf{H}^I(0, \mathbf{E}_r) + \mathbf{n}_{12} \times \mathbf{H}^{II} = -\mathbf{H}^I(\mathbf{j}^I, 0) \times \mathbf{n}_{21}. \qquad (3.23)$$

Equation (3.23) cannot be solved analytically for complicated electrodynamic systems, and one has to invoke numerical techniques. One of the most efficient solvers in this case is the Galerkin method [48], which is relatively simple and versatile. In addition, by averaging the boundary conditions in this approach one has the advantage that the energy is conserved in crossing the conventional boundary.

In accordance with the Galerkin method, we will expand the tangential electric field in terms of a complete set of plane vector orthonormal functions:

$$\mathbf{E}_r = \sum_r e_r \Xi_r, \qquad (3.24)$$

where e_r are unknown expansion coefficients.

Substituting expansion (3.24) into equation (3.23), then multiplying the latter by Ξ_q and integrating over the matching iris aperture S_s, we obtain the system of equations for the unknown coefficients e_r

$$\sum_r e_r \int_{S_s} \Xi_q \cdot \left(\mathbf{n}_{21} \times \mathbf{H}^I(0, \Xi_r)\right) ds \mp \int_{S_s} \Xi_q \cdot (\mathbf{n}_{12} \times \mathbf{H}^{II}) ds$$

$$= -\int_{S_s} \Xi_q \cdot \left(\mathbf{H}^I(\mathbf{j}^I, 0) \times \mathbf{n}_{21}\right) ds, \quad (3.25)$$

where $r = 1, ..., R$ and $q = 1, ..., R$.

The order of the system (3.25) is controlled by the number of coupling irises between the accelerating structure and the waveguide input sections, as well as by the number of the vector functions Ξ_r that are retained in the expansion of $\mathbf{E}_r$ at each matching iris.

The complete set of the vector orthonormal functions Ξ_r can be constructed using scalar functions ψ_e and ψ_h that satisfy the equation

$$\nabla^2 \psi_{e,h} + \kappa_{e,h}^2 \psi_{e,h} = 0$$

at the matching iris aperture and simultaneously satisfy the boundary conditions

$$\frac{\partial \psi_h}{\partial \mathbf{n}} = 0, \quad \psi_e = 0$$

on the contour of the iris aperture. Here, $\mathbf{n}$ is the normal to the contour.

The tangent electric field strength $\mathbf{E}_r$ may be written as the sum

$$\mathbf{E}_r = \sum_r e_r \Xi_r = \sum_h e_h \Xi_h + \sum_e e_e \Xi_e,$$

where $\Xi_h = \nabla \psi_h \times \mathbf{n}$ and $\Xi_l = \nabla \psi_l$ ($\mathbf{n}$ is the normal to the matching iris aperture).

The vector functions Ξ_r obey the orthonormalization conditions

$$\int_{S_s} \Xi_r \Xi_q \, ds = \delta_{rq} = \begin{cases} 0, & r \neq q, \\ 1, & r = q. \end{cases}$$

For practical uses of the system of equations (3.25), it is necessary to specify the integrands.

Let us consider the first term on the left-hand side of (3.25). It contains the quantity $\mathbf{H}^I(0; \Xi_r)$, which represents the field excited in the waveguide from its end by the tangent electric field Ξ_r at the coupling iris. To specify this term, we note that if an incident wave (for example, a wave from the generator) relates to a certain TM or TE type, then this type holds in the whole waveguide system including conventional boundaries. Let a TE-wave be propagating along the feeding waveguide from the generator. In this case, we will represent the desired field in the first partial domain as an expansion in terms of TE-waves:

$$
\left.
\begin{aligned}
\mathbf{E}^I &= \sum_h U_{ht}\mathcal{E}_{ht} = \sum_h U_h\mathcal{E}_h, \\
\mathbf{H}^I &= \sum_h I_{hz}\mathcal{H}_{hz} + \sum_h I_{ht}\mathcal{H}_{ht} = \sum_h I_h\mathcal{H}_h,
\end{aligned}
\right\}
\tag{3.26}
$$

where the indices t and z indicate that the corresponding vector relates to transverse and longitudinal vectors, respectively.

The functions $\mathcal{E}$ and $\mathcal{H}$ represent a set of vector orthonormal functions satisfying homogeneous boundary conditions on the lateral surface of the regular waveguide and U_h and I_h are the unknown expansion coefficients. The vector functions $\mathcal{E}_h$ and $\mathcal{H}_h$ are constructed similarly to the vector functions Ξ_r, using the scalar functions ψ_e and ψ_h.

In view of (3.26), the integral appearing in the first term on the left-hand side of (3.25) may be written as

$$
\int\limits_{S_s} \Xi_q \cdot \left(\mathbf{n}_{21} \times \mathbf{H}^I(0,\Xi_r)\right) ds = \sum_h I_h(0,\Xi_r) \int\limits_{S_s} \Xi_q \cdot \left(\mathbf{n}_{21} \times \mathcal{H}\right)_h ds . \tag{3.27}
$$

The unknown expansion coefficients $I_h(0, \Xi_r)$ are determined through $U_h(0, \Xi_r)$, which in turn are found from the orthonormalization condition for $\mathcal{E}_h$:

$$U_h(0,\Xi_r) = \int_{S_s} \Xi_r \cdot \mathcal{E}_h \, ds,$$

$$I_h(0,\Xi_r) = \frac{1}{Z_{0h}} \int_{S_s} \Xi_r \cdot \mathcal{E}_h \, ds, \qquad (3.28)$$

where Z_{0h} is the wave impedance for a wave with index h.

Substituting (3.28) into (3.27) and taking into account that $\mathcal{H}_h \times \mathbf{n}_{21} = \mathcal{E}_h$, we obtain for the first term of (3.25):

$$\sum_r e_r \int_{S_s} \Xi_q \cdot \left(\mathbf{H}^I(0,\Xi_r) \times \mathbf{n}_{21} \right) ds$$

$$= \sum_r e_r \sum_h \frac{1}{Z_{0h}} \int_{S_s} \Xi_q \cdot \mathcal{E}_h \, ds \int_{S_s} \Xi_r \cdot \mathcal{E}_h \, ds. \qquad (3.29)$$

Let us consider the right-hand side of equations (3.25), where the field excited by the extraneous source $\mathbf{j}^I$ in the waveguide short-circuited at the end ($\mathbf{E}_r = 0$) is uknown. Using the above procedure once again, we may write

$$\int_{S_s} \Xi_q \cdot \left(\mathbf{H}^I(\mathbf{j}^I,0) \times \mathbf{n}_{21} \right) ds = \sum_h I_h(\mathbf{j}^I,0) \int_{S_s} \Xi_q \cdot \mathcal{E}_h \, ds , \qquad (3.30)$$

For determining the coefficients $I_h(\mathbf{j}^I, 0)$, we will assume that the power P_h carried to the end of the waveguide by a wave with index h is known. According to the Poynting theorem, we have

$$P_h = \frac{1}{2} \int_{S_{wg}} (\mathbf{E}_h \times \mathbf{H}_h) Z_0 \, ds ,$$

where S_{wg} is the cross section of the waveguide.

Taking into consideration the orthonormalization conditions for the vector functions $\mathcal{E}_h$ and $\mathcal{H}_h$ and the relationship between the expansion coefficients I_h and U_h, we obtain

$$P_h = \frac{1}{2} Z_{0h} I_h^2 \int\limits_{S_{wg}} (\mathcal{E}_h \times \mathcal{H}_h) Z_0 \, ds = \frac{1}{2} Z_{0h} I_h^2 \int\limits_{S_{wg}} \mathcal{E}_h^2 \, ds = \frac{1}{2} Z_{0h} I_h^2 . \quad (3.31)$$

Thus, the amplitude coefficient of the wave with index h incident from the generator may be determined as

$$I_h = \sqrt{\frac{2P_h}{Z_{0h}}} = \sqrt{2P_h Y_{0h}} \, ,$$

where $Y_{0h} = 1/Z_{0h}$.

Since the waveguide is short-circuited ($\mathbf{E}_\tau = 0$) and thus standing waves occur in the waveguide (attenuation in the waveguide walls is neglected), relation (3.30) may be written in the form

$$\int\limits_{S_s} \Xi_q \cdot \left(\mathbf{H}^I(\mathbf{j}',0) \times \mathbf{n}_{21} \right) ds = \sum_h 2\sqrt{2P_h Y_{0h}} \int\limits_{S_s} \Xi_q \cdot \mathcal{E}_h \, ds . \quad (3.32)$$

Consider the second term on the left-hand side of (3.25). Here, the unknown quantity is the magnetic field in a cavity which is coupled with the feeding waveguide and excited by the component $\mathbf{E}_\tau$ at the coupling iris aperture. We will represent this field as an expansion in terms of the vector eigenfunctions of the cavity:

$$\mathbf{H}^{II}(0, \mathbf{E}_\tau) = \sum_m I_{mp}(0, \mathbf{E}_\tau) \mathbf{H}_{mp} , \quad (3.33)$$

where the index p relates to cavity p of the structure. In view of (3.33), we obtain

$$\int\limits_{S_s} \Xi_q \cdot (\mathbf{H}^{II} \times \mathbf{n}_{12}) ds = \sum_m I_{mp}(0, \mathbf{E}_\tau) \int\limits_{S_s} \Xi_q \cdot (\mathbf{H}_{mp} \times \mathbf{n}_{12}) ds . \quad (3.34)$$

Thus, the system of equations (3.25) for the unknown coefficients e_r may be written in the form:

$$\sum_r e_r \sum_h \frac{1}{Z_{0h}} \int_{S_s} \Xi_q \cdot \mathcal{E}_h \, ds \int_{S_s} \Xi_r \cdot \mathcal{E}_h \, ds$$

$$+ \sum_m I_{mp}(0, \mathbf{E}_\tau) \int_{S_s} \Xi_q \cdot (\mathbf{H}_{mp} \times \mathbf{n}_{12}) ds$$

$$= - \sum_h 2\sqrt{2P_h Y_{0h}} \int_{S_s} \Xi_q \cdot \mathcal{E}_h \, ds. \qquad (3.35)$$

The order of the system (3.35) is lower than the number of unknowns, because the coefficients I_{mp} are to be determined along with the expansion coefficients e_r.

To derive the complete system of equations for the unknown expansion coefficients e_r and I_{mp}, it is necessary to solve the problem of exciting a cell p of the accelerating structure by the tangent electric field specified at the matching iris aperture (coupling window). For this purpose, we will consider the second equation in (2.45)

$$\left[k_{mp}^2 - k^2 + (j-1)\frac{k k_{mp}}{Q_{mp}} \right] I_{mp} = -j\omega\varepsilon_0 \int_{S_s} (\mathbf{E}_p \times \mathbf{H}_{mp}) \cdot \mathbf{n} \, ds \, .$$

The integral over S_s contains both the integral over the coupling slots between adjacent resonators and the integral over the coupling window of the rf input coupler. Taking this point into consideration, we may write

$$\left[k_{mp}^2 - k^2 + (j-1)\frac{k k_{mp}}{Q_{mp}} \right] I_{mp}$$

$$= -j\omega\varepsilon_0 \int_{S_{sl}} (\mathbf{E}_p \times \mathbf{H}_{mp}) \cdot \mathbf{n} \, ds - j\omega\varepsilon_0 \int_{S_{cw}} (\mathbf{E}_p \times \mathbf{H}_{mp}) \cdot \mathbf{n} \, ds, \qquad (3.36)$$

where S_{sl} is the surface of the coupling slots with other resonators and S_{cw} is the coupling window surface.

Substituting the expansion of the tangent electric field (3.24) into the integral over the coupling window, using relation (2.46), and combining the systems of equations (3.35) and (3.36), we obtain the complete system of equations for the expansion coefficients of the field configuration in the accelerating structure and for the expansion coefficients of the field at the surface of the coupler window:

$$
\left[k_{mp}^2 - k^2 + (j-1)\frac{kk_{mp}}{Q_{mp}} \right] I_{mp} + j\omega\varepsilon_0 \sum_m I_{mp} C_m^p
$$

$$
+ j\omega\varepsilon_0 \sum_m I_{mk} C_m^k + j\omega\varepsilon_0 \sum_r e_r \int_{S_{cw}} (\Xi_r \times \mathbf{H}_{mp}) \cdot \mathbf{n} ds = 0,
$$

$$
\sum_m I_{mp} \int_{S_{cw}} \Xi_q \cdot (\mathbf{H}_{mp} \times \mathbf{n}_{12}) ds
$$

$$
+ \sum_r e_r \sum_h \frac{1}{Z_{0h}} \int_{S_{cw}} \Xi_q \cdot \mathcal{E}_h \, ds \int_{S_{cw}} \Xi_r \cdot \mathcal{E}_h \, ds
$$

$$
= -\sum_h 2\sqrt{2P_h Y_{0h}} \int_{S_{cw}} \Xi_q \cdot \mathcal{E}_h \, ds.
$$

$$(3.37)$$

The order of the system of equations (3.37) is $M \times P + R \times NV$, where M is the number of the vector eigenfunctions retained in the expansion of fields in the cavities of the accelerating system, P is the number of resonators in the accelerating system, R is the number of plane vector functions retained in the expansion of $\mathbf{E}_r$ in the matching iris aperture, and NV is the number of coupling windows.

The solution of system (3.37) makes it possible to determine the dynamic range and external parameters of the accelerating structure: the distribution of amplitudes and phases over cavities that takes into account the connected waveguides, the standing-wave ratios in the waveguides.

Thus, the system of equations (3.37) can be used in comprehensive numerical studies of complex cavity structures with allowance for connected waveguide sections.

3.4 Coupler: Cavity-Analogue Analysis

Here we determine the input impedance of a cylindrical cavity excited in TE_{0n0} modes in the case when the cavity is coupled with a rectangular waveguide through an inductive iris coincident with the side surface of the cavity ($r = R$). This configuration is typical for a standing wave electron linac in the range $\lambda \le 0.3$ m.

As a cavity-analogue, we take a cavity formed from a short-circuited rectangular waveguide and an inductive iris and excited in TE_{101} modes [94]. An analytic solution known for this cavity-analogue will be reproduced with reference to the scheme shown in Fig. 3.7. The amplitudes of reflected waves b_1 and b_2 are determined from the matrix equation

$$\begin{pmatrix} -\sqrt{1-k^2} & jk \\ jk & -\sqrt{1-k^2} \end{pmatrix} \begin{pmatrix} a_1 \\ -b_2 e^{-j(\alpha_t + 2\varphi)} \end{pmatrix} = \begin{pmatrix} b_1 \\ b_2 \end{pmatrix}, \qquad (3.38)$$

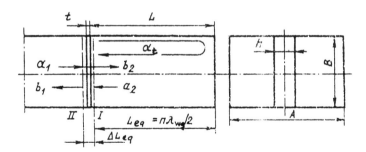

Figure 3.7. To illustrate the calculations of the input impedance of a cavity formed by a section of a short-circuited rectangular waveguide and an inductive iris (I and II are the planes of equivalent representation).

where a_1 is the amplitude of an incident wave, b_1 is the amplitude of the reflected wave entering the cavity, and α_t is the total attenuation due to a double passage through the cavity. The quantity k is related to the unitary scattering matrix of the input coupler S by

$$S_{11} = S_{22} = -\sqrt{1-k^2} \, .$$

In view of the equality

$$k = 2x \big/ \sqrt{4x^2 + 1} \, ,$$

where x is the iris impedance normalized by the rectangular waveguide impedance. The solution of equations (3.38), accurate to the first order terms, has the form

$$Z_{\text{in}} = \frac{1+\Gamma}{1-\Gamma} = \frac{2x^2 - x^2 \alpha_t}{\alpha_t + x^2 \alpha_t} \approx \frac{2x^2}{\alpha_t} \, . \tag{3.39}$$

When deriving (3.39), we take into account that $\alpha_t \ll 1$ and that

$$\Gamma = \frac{b_1}{a_1} = -\left(\sqrt{1-k^2} - \exp(-\alpha_t)\right)\!\left(1 - \sqrt{1-k^2}\,\exp(-\alpha_t)\right).$$

at resonance ($\varphi = \pi n$). The attenuation α_t is related to the cavity Q-factor as follows:

$$\alpha_t = \frac{\pi p}{Q}\left(\frac{\lambda_{\text{wg}}}{\lambda_0}\right)^2 . \tag{3.40}$$

Here, $p = 1, 2, 3, \ldots$ is the number of half-waves along the cavity.

For an inductive iris in a rectangular waveguide with $h/A < 0.3$, we have [11]

$$x = -\frac{A}{\lambda_{\text{wg}}} \tan^2 \frac{\pi(h-t)}{2A} , \tag{3.41}$$

where A is the width of the wide waveguide wall, h is the width of the iris aperture, and t is the thickness of the iris.

Thus, the input impedance of a cavity (excited in the TE_{10} mode) constituted by an inductive iris and a short-circuited waveguide may be written in the form

$$Z_{in} = \frac{2Q^\square}{\pi p}\left(\frac{\lambda_0}{\lambda_{wg}}\right)^2\left(\frac{A}{\lambda_{wg}}\right)^2 \tan^4 \frac{\pi(h-t)}{2A}, \qquad (3.42)$$

where $Q^\square$ is the Q factor of a prismatic cavity. It is obvious that the input impedance of the considered system will not change if the prismatic cavity is replaced by any other geometry provided that

$$H_\tau^\square(x,y) = H_\tau(x,y), \qquad (3.43)$$

where $H_\tau^\square(x,y)$ is the tangential component of the magnetic field at the coupling window from the side of the prismatic cavity and $H_\tau(x, y)$ is the same component from the side of the equivalent cavity. When restricted to the case of the fundamental wave, condition (3.43) switches to

$$Q^\square \xi_{M\square S}^2 = Q\xi_{MS}^2, \qquad (3.44)$$

where $\xi_{M\square S}$ and ξ_{MS} are the normalized parameters $(H/\sqrt{PQ})$ of the magnetic field at the metallic surface of the matching iris of the prismatic and equivalent cavity, respectively. If the prismatic cavity is replaced by a cylindrical cavity resonant in TM_{0n0} modes, then

$$Q\xi_{MS}^2 = \frac{H^2(S)}{P} = \frac{4Q}{v_{0n}^2 Z_0 \lambda_0 L}$$

and

$$Q^\square \xi_{M\square S}^2 = \frac{H_0^2(S)}{P} = \frac{8Q^\square}{\pi Z_0 AB}\left(\frac{\lambda_0}{\lambda_{wg}}\right)^3 \frac{1}{n}\cos^2\frac{\pi x}{A}, \qquad (3.45)$$

where B is the height of the rectangular waveguide.

Using (3.39), (3.42), and (3.44), we obtain the normalized input impedance of the cylindrical cavity resonant in the TM_{0n0} mode that is coupled with a rectangular waveguide by an inductive iris:

$$Z_{in} = \frac{Q}{v_{0n}^2} \frac{(A/\lambda_0)^3}{L/\lambda_0} \frac{B/\lambda_0}{\lambda_{wg}/\lambda_0} \tan^4 \frac{\pi(h-t)}{2A}. \qquad (3.46)$$

If the width of the iris aperture satisfies the condition $h/A \geq 0.3$, then more accurate formulas will be

$$Z_{in} = \frac{Q}{v_{0n}^2} \frac{A}{\lambda_0} \frac{\lambda_{wg}}{\lambda_0} \frac{B}{L} \left\{ 4 + \left(\frac{\lambda_{wg}}{A}\right)^2 \cot^4 \frac{\pi(h-t)}{2A} \right\}^{-1} \frac{\sin^2[\pi h/(2A)]}{[\pi h/(2A)]^2},$$
$$(3.47a)$$

$$Z_{in} \approx \frac{QAB\lambda_{wg}Z_0\xi_{MS}^2}{4\lambda_0} \left\{ 4 + \left(\frac{\lambda_{wg}}{A}\right)^2 \cot^4 \frac{\pi(h-t)}{2A} \right\}^{-1} \frac{\sin^2[\pi h/(2A)]}{[\pi h/(2A)]^2},$$
$$(3.47b)$$

where the magnetic field parameter $\xi_{M\square S}$ is averaged over the generatrix of the coupling window and the second-order terms are taken into account in (3.39). Formula (3.47b) is convenient to use in the case when an experimental value of ξ_{MS} is specified.

The effect of the inductive iris on the cavity frequency can be estimated from the known relations for the equivalent length of the prismatic cavity (see Fig. 3.7):

$$\Delta L_{eq} = \frac{\lambda_{wg}}{4\pi} \arctan(2x). \qquad (3.48)$$

The deviation of the frequency of the prismatic cavity excited in TE_{101}-modes is related to its length variation by the known formula

$$\frac{\Delta f}{f_0} = \frac{f_1 - f_0}{f_0} = -\left(\frac{\lambda_0}{\lambda_{wg}}\right)^2 \frac{\Delta L_{eq}}{L_{eq}}, \qquad (3.49)$$

where $L_{eq} \approx \lambda_{wg}/2$ and $\lambda_{wg} = \lambda_0[1 - (\lambda_0/2A)^2]^{-1/2}$.

By the theorem of small perturbations, we can also write the relative deviation in the form

$$\Delta f/f_0 = \mathcal{H}_i^m H_S^2/W,$$

where $\mathcal{H}_i^m$ is the form factor of the iris aperture with respect to the magnetic field. Using the relation

$$\frac{H_S^2}{W} = \frac{8\lambda_0^2 c}{Z_0 ABL\lambda_{wg}^2},$$

we obtain the form factor of the coupling window in a uniform magnetic field

$$\mathcal{H}_i^m = \frac{Z_0 AB\Delta L_{eq}}{8c}\left[\frac{\sin \pi h/(2A)}{\pi h/(2A)}\right]^{-2}. \tag{3.50}$$

When the cylindrical cavity resonant in TM$_{0n0}$ modes is excited by an inductive iris made in its side surface, from (3.48), (3.50), and (3.41) we obtain

$$\frac{\Delta f}{f_0} \approx \mathcal{H}_i^m \frac{H_S^2}{W} = -\frac{A^2 BZ_0\xi_{MS}^2}{8\lambda_0}\tan^2\frac{\pi(h-t)}{2A}$$

$$= -\frac{A^2 B}{2v_{0n}^2\lambda_0^2 L}\tan^2\frac{\pi(h-t)}{2A}, \tag{3.51a}$$

and

$$\frac{\Delta f}{f_0} = \frac{AB\lambda_{wg}Z_0\xi_{MS}^2}{16\lambda_0}\arctan\left[\frac{2A}{\lambda_{wg}}\tan^2\frac{\pi(h-t)}{2A}\right]\frac{[\pi h/(2A)]^2}{\sin^2[\pi h/(2A)]}$$

$$= -\frac{AB\lambda_{wg}}{4v_{0n}^2\lambda_0^2 L}\frac{[\pi h/(2A)]^2}{\sin^2[\pi h/(2A)]}\arctan\left[\frac{2A}{\lambda_{wg}}\tan^2\frac{\pi(h-t)}{2A}\right]. \tag{3.51b}$$

for $\pi(h - t)/2A < 0.4$ and $\pi(h - t)/2A \geq 0.4$, respectively.

Relations (3.46), (3.47), and (3.51) were extensively tested by experimental data. The most complete verification was conducted during the tunup of the accelerating subsections of the RELUS-1 setup [95]. The following input data were used for calculations: $\lambda_0 = 107.2$ mm, $A \times B = 72 \times 34$ mm, $Q = 1.04 \times 10^4$, $t = R[1 - \sqrt{1 - (h/R)^2}$, $L = N(D - 2t - L_c) = 7(53.6 - 8 - 4) = 292$ mm, and $R \approx 0.383\lambda_0 = 41.1$ mm. The frequency deviation for the input cell is calculated at $L = 41.6$ mm. Table 3.2 summarizes the calculated and measured results of this test.

Table 3.3 includes Z_{in} calculated in [92] by the structurally very complicated GROT program based on a numerical modeling. In nontrivial cases, when magnetic field parameter at the iris aperture is hard to estimate, experimental values of ξ_{MS} should be used. Such a situation occurred in the design of an output coupler for an ultrarelativistic klystron [96] whose toroidal cavity was excited in the TM_{020} mode and was shortened by a capacitance gap comparable in length with the other dimensions of the cavity. Here, the following initial data were obtained: $f_0 = 2980$ MHz, $\xi_M(r = R) = (0.2 + 0.01) \Omega^{-1/2}m^{-1}$, and $Q = 1.05 \times 10^4$. The cross

Table 3.2 Calculated and measured data for a coupler to a cavity excited in the TM_{010} mode

h, mm	t, mm	$Z_{in}^{(calc)}$	$Z_{in}^{(exp)}$	$Z_{in}^{(calc)}$ [92]	$-\Delta f^{(calc)}$ MHz	$-\Delta f^{(exp)}$ MHz
12	0.5	0.18	0.18	0.25	5.3	5.5
13	0.6	0.25	0.25	—	6.9	6.4
14	0.7	0.34	0.33	0.38	8.0	7.8
15	0.8	0.445	0.46	—	9.1	8.8
16	0.9	0.58	0.59	0.63	10.4	10.2
17	1.0	0.74	0.80	0.83	11.8	11.7
18	1.1	0.95	1.0	0.90	13.3	13.0
19	1.2	1.19	1.25	1.0	14.9	14.5
20	1.3	1.50	1.6	1.15	16.6	16.0
20.5	1.3	1.70	1.8	1.3	17.7	17.0

Table 3.3 Calculated and measured data for an rf power feeder to a cavity excited in the TM$_{020}$ mode

$h - t$, mm	$Z_{in}^{(calc)}$	$Z_{in}^{(exp)}$	$-\Delta f^{(calc)}$, MHz	$-\Delta f^{(exp)}$, MHz
2	0	0	0	0
11	0.13	0.15	0.6	1.0
17	0.75	0.79	1.6	2.0
19	1.22	1.20	2.0	2.5
23	2.7	2.7	3.1	4.0
30.5	8.3	8.1	6.3	7.0

section of the klystron waveguide was $A \times B = 72 \times 34$ mm. The experimental and theoretical results obtained with formulas (3.47b) and (3.51b) are summarized in Table 3.3

Tables 3.2 and 3.3 prove that the cavity-analogue method is a correct model for waveguide coupler design. A comparison of calculated data with experimental data obtained in the tuneup of input couplers of RELUS linacs [145] also proves the reliability and feasibility of this method.

3.5 Coaxial Feed Couplers

Loop couplers of accelerating structures with coaxial waveguides are widely used in meter and centimeter wave bands. For linear ion accelerators, this type of power feeding devices is the only possible solution because of their reasonable dimensions and electric power. Only fragmentary calculations of loop couplers are available in the literature [98, 99].

The voltage induced in the loop coupled with the magnetic field of a cavity is

$$U_1 = -\frac{d\Phi}{dt} = -\frac{d}{dt}(j\mu_0 S_1 H_1 \sin \omega_0 t) = -j\mu_0 S_1 \omega_0 H_1, \quad (3.52)$$

where S_1 is the loop area and H_1 is the mean magnetic field in the loop plane. The voltage U_g excited in the loop by a generator must counter the induced voltage and the self-induction emf of the loop:

$$U_g = j\omega_0\mu_0 S_1 H_1 + j\omega_0 L_1 I_1 , \qquad (3.53)$$

where I_1 is the amplitude of loop current and L_1 is the loop inductance that includes the reactive load due to the excitation of higher order modes (HOMs). The active power of the generator is related only to the first term in (3.53) and is equal to $P_0 = U_g I_g = \omega_0\mu_0 S_1 H_1 I_1$. Assuming the magnetic field to be uniform over the loop and taking into account that P_0 is the amplitude of power, we express the magnetic field through the magnetic field parameter $H_1^2 = P_0 \xi_M^2 Q/2$. From the latter two relations, we find $H_1 = \omega_0\mu_0 S_1 \xi_M^2 Q I_1/2$ to be substituted into (3.53) at $L_1 = 0$. As a result, we obtain the normalized input impedance of the loop

$$Z_{in} = \frac{R_{in}}{R_0} = \frac{U_g}{I_1 R_0} = \frac{Q}{2R_0}(\omega_0\mu_0 S_1 \xi_M)^2 , \qquad (3.54)$$

where R_0 is the wave impedance of the coaxial waveguide coupled with the loop. In the case of an H-type resonator with length L and radius R, the loop area can be calculated by the formula

$$S_1 = \frac{1.32R}{\omega_0\mu_0}\left(\frac{2R_0 Z_{in}}{Q}\right)^{1/2}\left[\frac{Z_0 L(1 + 2.91 R^2/L^2)}{\lambda_0}\right]^{1/2} . \qquad (3.55)$$

where relation (5.16) is taken into account. The importance of inclusion of the inductance of a coupling loop can be shown by the following example. Suppose the resonance is indicated by a minimum of a reflected wave in a directional coupler when the input impedance is purely active. In this case, the loop inductance is compensated at the generator frequency that is determined from the equation

$$\frac{2jQ}{Z_{in}}\frac{f_g - f_0}{f_0} + \frac{j\omega_0 L_1}{R_0} = 0 , \qquad (3.56)$$

and the generator will be detuned by

$$\Delta f = f_g - f_0 = -\pi f_0^2 L_1 Z_{in}/(QR_0) .$$

The equation of a Q-curve [11] enables one to find the relationship between the actual E_a and calculated E_c strengths of the electric field at the center of a gap:

$$\frac{E_a}{E_c} = \left[1 + \left(\frac{2\pi f_0 L_1 Z_{in}}{R_0} \right)^2 \right]^{-1/2} . \qquad (3.57)$$

For the numerical values for the Uragan-2 accelerator [47]: $L = 0.54$ m, $\lambda_0 = 2$ m, $R = 0.163$ m, $Z_{in} = 1.2$, $R_0 = 50\ \Omega$, and $Q = 4.7 \times 10^3$, we have $S_1 \approx 4$ cm^2. The inductance of a square loop with side A_1 is

$$L_1 = \frac{2\mu_0 A_1}{\pi} \left(\ln \frac{A}{r_0} - 0.22 \right), \qquad (3.58)$$

where r_0 is the radius of wire from which the loop is made. At $r_0 = 1$ mm and $A_1 = 20$ mm, we find that $L_1 = 4.5 \times 10^{-8}$ H. Substituting this value of inductance into (3.57), we obtain $E_a/E_c = 0.7$. The setup proves to be inoperative, because the generator power must be doubled to reach the rated value of E.

Relations (3.55), (3.57), and (3.58) lead us to a criterion for estimating the cross section S_{cs} of the square loop conductor

$$S_{cs} \geq 2S_1 \exp \left(-\frac{R_0 \sqrt{1 - \eta^2}}{2 f_0 \mu_0 \eta \sqrt{S_1}} \right), \qquad (3.59)$$

where $\eta = P_0/P_g = R_0 [R_0^2 + (\omega L_1)^2]^{-1/2}$ is the efficiency of the reactively loaded generator. If $\eta \geq 0.9$, then $S_{cs} \geq 32$ mm^2 in the considered example of the Uragan-2 accelerator. A coupling loop with an area of 4 cm^2 should be made, e.g., of 10×4 mm copper bus.

We should also mention here the design of couplers with resonant mockups of DLWs and BPSs designed for experimental studies on a semi-automated wideband cavity test systems. In these systems, a cavity is connected in a through circuit with symmetric coupling slots and the power attenuation in the cavity is

$$A = 20\log\frac{2Z_{in}}{1+2Z_{in}}. \qquad (3.60)$$

In the interval $R \le R_1 \le 0.6R$, where R_1 is the radial coordinate of a loop, the magnetic field parameter is described by relation (3.45) with an error within 10%. Substituting (3.45) into (3.54) and using (3.60), we obtain a formula to calculate the transition power attenuation through a resonant mockup of a DLW or BPS with symmetric loop couplers in a configuration with the plane of loops perpendicular to the H-component of the field in the mockup:

$$A \approx 20\log\left[\frac{16\pi^2 Z_0 Q}{\lambda_0^3 v_0^2 R_0 L_{eq}}(S_1 + S_0)^2\right], \qquad (3.61)$$

where $L_{eq} = L$ for a DLW and $L_{eq} = N(D - 2t - L_c)$ for a BPS.

The physical meaning of the quantities S_0 and S_1 in (3.61) is clarified in Fig. 3.8, which shows a controlled loop coupler

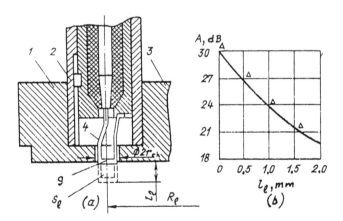

Figure 3.8. Loop coupler with a resonance mockup of a DLW or BPS: (a) sectional view: *1*, flange; *2*, directrix; *3*, loop holder; *4*, loop; (b) dependence of the transition attenuation on the location of coupling loops in the reference cavity: $\lambda_0 = 0.105$ m, $L = 0.11$ m, $Q = 1.2 \cdot 10^4$, $r_{ch} = 2$ mm, $h = 2.4$ mm; measured data are shown by triangles.

mounted on the end wall of the cavity: S_0 is the area of a part of the loop inserted into the cavity to a height l_1. The initial area of the loop is determined from the magnetic flux through the exposed loop portion (coupling channel) at $l_1 = 0$. The coupling channel is excited by the magnetic field of the cavity in TE_{11} modes beyond cutoff.

Equating the azimuthal components of the magnetic field in the cavity at $r = R_1$ and in the beyond-cutoff channel at the point g indicated in Fig. 3.8a, we obtain in the zeroth approximation

$$S_0 \approx \frac{2}{k_{cr}} \int_0^\infty \int_0^{h/2} \frac{J_1(\mu_{11}r/r_{ch})}{r} \exp(-k_{cr}z)\,dz\,dr \approx \frac{r_{ch}h}{2\mu_{11}}. \qquad (3.62)$$

Figure 3.8b displays the transition power attenuation calculated by formula (3.62) for the typical case of measurements with a reference cavity excited in the TM_{010} mode. It also shows measured data, which verify the calculation.

Pin couplers are most frequently used to measure the natural frequencies of BPS cells. A design of such a coupler is shown in

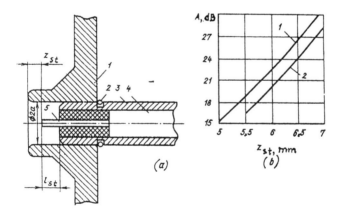

Figure 3.9. Pin couplers with BPS cells: (a) design: *1*, end wall of a cell; *2*, lock; *3*, pipe; *4*, coaxial cable; *5*, pin; (b) the transition attenuation as a function of pin position for an accelerating cell at $\lambda_0 = 0.1$ m, $a = 5$ mm, $l_{st} = 1$ mm, $\zeta_0 = 500\ \Omega^{1/2}/$m, and $R_0 = 50\ \Omega$ (curve *1*) and for a coupling cell at $Q = 2.5 \times 10^3$ and $L_c = 4$ mm (curve *2*).

Fig. 3.9a. A pipe *3* that bears a pin antenna *5* is fed into the beam channel of an accelerating system and coupled with the cells through the beyond-cutoff portion of drift tubes of length z_{st}. The power attenuation of this structure can be estimated as [98]:

$$Z_{in} = (E_{st} l_{eff})^2 / (2R_0 P), \qquad (3.63)$$

where E_{st} is the electric field along the stub and l_{eff} is the effective height of the antenna, which is equal to $l_{eff} = l_{st}/2$ for a linear distribution of current in the stub. We write the electric field on the axis of BPS cells in the form of relation (3.3) and assume that $E_{st} = E_0 \exp[-v_{01} z_{st}/a]$. As a result, in view of (3.60), we obtain from (3.63) for $Z_{in} < 0.1$

$$\left. \begin{aligned} A_a &\approx 20 \log \left[\frac{0.64 Z_0 L_{eq} Q l_{st}^2}{(d_g + 2a/v_{01})^2 \lambda_0 R_0} \exp\left(-\frac{2v_{01}}{a} z_{st} \right) \right] \\ &\approx 20 \log \left[\frac{\xi_{ef}^2 \exp(-2v_{01} z_{st}/a) Q l_{st}^2}{4 R_0} \right], \\[2em] A_c &\approx 20 \log \left[\frac{0.64 Z_0 Q l_{st}^2}{L_c \lambda_0 R_0} \exp\left(-\frac{2v_{01}}{a} z_{st} \right) \right]. \end{aligned} \right\} \qquad (3.64)$$

The second estimate in (3.64) relates to cylindrical coupling cells. The transition attenuation of an accelerating cell of a BPS is calculated by (3.64) and is plotted in Fig. 3.9b for typical values of cell and coupler parameters. The plot in Fig. 3.9b and relation (3.63) are for engineering estimations ($\delta A = \pm 3$ dB). This limitation is due to the fact that, in a rigorous formulation, the effective height of the pin must be calculated by including both its capacitance to nearby conducting surfaces and HOMs in the region of the stub. In practice the defects of the qualitative solution are corrected by choosing the stub coordinate (z_{st} in Fig. 3.9) during the calibration of the transition attenuation of the cavity on a semi-automatic test bench (see Section 6.2).

4

E– TYPE
ACCELERATING
STRUCTURES

4.1 Disk Loaded Waveguide

Disk loaded waveguides (DLWs) are widely used in different linacs. At the present time, they are treated as accelerating structures for linear colliders.

The purpose of a DLW design is to determine the DLW geometrical dimensions which provide for the required characteristics of the accelerated beam. Solving this problem, the designer calculates the particle dynamics in the electromagnetic field whose amplitude and phase velocity along the waveguide are functions of the DLW geometry. The dependence of the DLW dynamic range on the geometrical dimensions may be found in our handbooks [3, 4] where the reference data are represented in a versatile form excluding the dependence of all the characteristics on the operating frequency. The range of the characteristic parameters compiled in these texts has been chosen so as to cover the needs of the current and future designs. The high accuracy of these parameters was achieved with specially developed equipment and methods of experimental investigation of cavities tested in different DLW technologies. The data are presented in a graphical form for easy engineering estimations and in the form of detailed tables with a interpolation of consecutive data.

To be more specific, the handbook presents the following electrodynamic characteristics.

1. Dispersion relation in the form

$$a/\lambda_n = f(a/b, \beta_{\text{ph } n})$$

worked out for t/λ_n = constant, θ = constant, $R_\text{r} = t/2$, and without rounding. Here, a is the disk aperture radius, b is the internal DLW radius, t is the disk thickness, and R_r is the radius of the disk aperture rounding.

The resonance frequency of arbitrary mode can be evaluated from plots in the form

$$C_l/f_n = F(a/b, \beta_{\text{ph } n})$$

presented for t/λ_n = constant and θ = constant. Here, C_l is a coefficient of expansion of the dispersion function in a Fourier series.

2. Frequency sensitivity function

$$\frac{1}{f_n^2}\frac{\partial f_n}{\partial q_i} = F(a/b, \beta_{\text{ph } n})$$

worked out for t/λ_n = constant and θ = constant; q_i is any dimension of a DLW.

3. Group velocity

$$\beta_{\text{gr } n} = F(a/b, \beta_{\text{ph } n})$$

worked out for t/λ_n = constant and θ = constant.

4. Derivatives of the phase velocity with respect to the DLW dimensions

$$\lambda_n \frac{\partial \beta_{ph}}{\partial q_i} = F(a/b, \beta_{ph\,n})$$

worked out for t/λ_n = constant and θ = constant.

5. Axial electric component of the traveling wave (normalized amplitude of the main harmonic)

$$\xi_{e,\,TW} = \frac{E_0 \lambda_n}{\sqrt{P}} = F(a/b, \beta_{ph\,n})$$

worked out for t/λ_n = constant and θ = constant.

The quantity $\xi_{e,TW}$ is related to the normalized amplitude of the main harmonic $\xi_{e,1}$ of the electrical field on the axis of a resonance DLW mockup of length L by the formula

$$\xi_{e,\,TW} = \sqrt{\frac{\pi \lambda_n L}{\beta_{ph\,n}}} \, \xi_{e,\,1} \tag{4.1}$$

The quantity $\xi_{e,1}$ is determined by resolving the dependence

$$\xi_e = E_z(z) \big/ \sqrt{PQ} \tag{4.2}$$

for the main harmonic.

6. Shunt impedance per unit length

$$\lambda_n^{1/2} r_{sh} = F(a/b, \beta_{ph\,n})$$

worked out for t/λ_n = constant and θ = constant.

The wave impedance

$$\lambda_n \, r_{sh}/Q = F(a/b, \beta_{ph\,n})$$

worked out for t/λ_n = constant and θ = constant.

The overvoltage coefficient

$$k_{ov} = F(a/b, \beta_{ph\,n})$$

worked out for $t/\lambda_n = $ constant and $\theta = $ constant.

7. Attenuation constant

$$\alpha \lambda_n^{3/2} = F(a/b, \beta_{ph\,n})$$

worked out for $t/\lambda_n = $ constant and $\theta = $ constant.

The Q factor

$$Q \lambda_n^{-1/2} = F(a/b, \beta_{ph\,n})$$

worked out for $t/\lambda_n = $ constant and $\theta = $ constant.

We illustrate the range of these reference data by choosing the dimension of a DLW for a high-current linac and a structure for a 2 × 1 TeV linear electron-positron collider.

The design of a high-current linac involves a number of separate problems. We consider those connected with the choice of an accelerating system that provides the required energy and average beam power for given rf power parameters.

We write the expressions for the energy increment in volts and for the efficiency of accelerating structures with constant impedance [100]

$$\frac{U}{\sqrt{r_{sh}LP_0}} = \sqrt{\frac{2}{\tau}}(1 - e^{-\tau}) - I_0 \sqrt{\frac{r_{sh}L}{P_0}}\left(1 - \frac{(1 - e^{-\tau})}{\tau}\right), \qquad (4.3)$$

$$\eta = \sqrt{\frac{r_{sh}L}{P_0}}\sqrt{\frac{2}{\tau}}(1 - e^{-\tau})I_0 - \frac{r_{sh}L}{P_0}\left(1 - \frac{1 - e^{-\tau}}{\tau}\right)I_0^2, \qquad (4.4)$$

and constant gradient

$$\frac{U}{\sqrt{r_{sh}LP_0}} = \sqrt{1 - e^{-2\tau}} - I_0\frac{r_{sh}L}{P_0}\left(\frac{1}{2} - \frac{\tau e^{-2\tau}}{1 - e^{-2\tau}}\right), \qquad (4.5)$$

$$\eta = \sqrt{\frac{r_{sh}L}{P_0}}\sqrt{1-e^{-2\tau}}I_0 - \frac{r_{sh}L}{P_0}\left(\frac{1}{2} - \frac{\tau e^{-2\tau}}{1-e^{-2\tau}}\right)I_0^2 . \qquad (4.6)$$

Here, P_0 is the power flux at the origin of the structure ($z = 0$), I_0 is the current of electrons, L is the length of the structure, $\tau = \alpha L$ is the attenuation parameter, $r_{sh} = E_0^2/(2\alpha P_0)$ is the shunt impedance per unit length, and E_0 is the amplitude of the longitudinal accelerating component of the electric field.

Expressions (4.3)–(4.6) are written on the assumption that the phase of the bunch of electrons with respect to the rf field maximum is zero.

An analysis of expressions (4.4) and (4.6) indicates that, to obtain a high efficiency, the loss of rf power in the wall must be made small compared to the power expended for beam acceleration. Therefore, the attenuation parameter must be small ($\tau < 0.25$). It is worth noting, that with such values of τ, the efficiencies calculated by formulas (4.4) and (4.6) are almost identical, therefore, without any loss of generality, we may confine our subsequent analysis to structures with constant impedance.

Using expressions (4.3) and (4.4) we write the current I_0 and efficiency in terms of the accelerating-structure parameters α, r_{sh}, beam energy U, and rf power P_0

$$I_0 = \frac{\sqrt{r_{sh}LP_0}\sqrt{\frac{2}{\tau}}(1-e^{-\tau}) - U}{r_{sh}L\left(1 - \frac{(1-e^{-\tau})}{\tau}\right)}, \qquad (4.7)$$

$$\eta = \frac{\alpha L}{\alpha L + e^{-\alpha L} - 1}\left\{\sqrt{\frac{2}{r_{sh}L^2 P_0 \alpha}}(1-e^{-\alpha L})\left[\sqrt{\frac{2r_{sh}P_0}{\alpha}}(1-e^{-\alpha L}) - U\right]\right.$$

$$\left. - \frac{1}{r_{sh}LP_0}\left[\sqrt{\frac{2r_{sh}P_0}{\alpha}}(1-e^{-\alpha L}) - U\right]^2\right\} \qquad (4.8)$$

An analysis of (4.8) indicates that the efficiency can be increased by decreasing the attenuation in the structure. However, this gain in efficiency requires a considerably longer structure.

The length of a linac can be reduced without having to considerably sacrifice the efficiency if the length is limited by the section where the rf power flux become attenuated to P_L. Then the efficiency may be estimated by the formula

$$\eta = 2\left[e^{-\tau} - \sqrt{\frac{P_0}{P_L}} - (\tau + e^{-\tau} - 1)\left(\frac{e^{-\tau} - \sqrt{P_L/P_0}}{e^{-\tau} - 1}\right)^2\right] \qquad (4.9)$$

and the section length determined from the equation

$$e^{-\tau} - \left(e^{-\tau} - \sqrt{\frac{P_L}{P_0}}\right)\left(1 + \frac{\tau}{e^{-\tau} - 1}\right) = 1 - U\sqrt{\frac{\alpha}{2P_0 r_{\text{sh}}}}. \qquad (4.10)$$

The respective data for the two accelerating structures are summarized in Table 4.1. Clearly, if the linac length is limited by the cross section where $P_L = P_0/e^2$, the efficiency will be reduced insignificantly. For this case, the expressions for L, U, and η take the form [1]

$$e^{-\tau} - \left(e^{-\tau} - 1/e\right)\left(1 + \frac{\tau}{e^{-\tau} - 1}\right) = 1 - U\sqrt{\frac{\alpha}{2P_0 r_{\text{sh}}}}, \qquad (4.11)$$

$$U = \sqrt{\frac{2P_0 r_{\text{sh}}}{\alpha}}\left[1 - e^{-\alpha L} - \frac{e^{-\alpha L}(\alpha L + e^{-\alpha L} - 1)(1 - e^{\alpha L - 1})}{1 - e^{-\alpha L}}\right], \qquad (4.12)$$

$$\eta = 2\left[e^{-\tau} - 1/e - (\tau + e^{-\tau} - 1)\left(\frac{e^{-\tau} - 1/e}{e^{-\tau} - 1}\right)^2\right]. \qquad (4.13)$$

Table 4.1 L [m] and η [%] for a constant-impedance DLW linac excited in the $2\pi/3$ mode at 915 MHz presented as functions of P_L/P_0

P_0	U	P_L/P_0									
		0.1		$1/e^2$		0.2		0.3		0.4	
MW	MV	L	η	L	η	L	η	L	η	L	η
5	8	4.57	80	4.41	71	4.15	69	3.87	59	3.67	48
10	10	4.03	81	3.90	77	3.67	70	3.46	60	3.24	50

Figure 4.1 shows U and η versus the length of the accelerating structure for three DLWs (labeled 1, 2, and 3), and a structure with conducting stubs (SCS) labeled by subscript 5. The electrodynamic characteristics of these structures are summarized in Table 4.2 for the $2\pi/3$ mode at 915 MHz and $\beta_{ph} = 1$. The DLW data have been borrowed from our handbook [4] and the SCS data from our paper on SW structures [101]. Calculations have been performed with formulas (4.11)–(4.13) for $P_0 = 10$ MW.

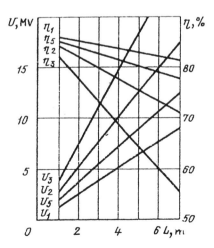

Figure 4.1 Dependence of U and η on the length of DLWs (subscripts 1, 2, and 3) and a structure with cross bars (subscript 5)

Table 4.2 Parameters of DLWs and an SCS at 915 MHz

Parameters	DLW$_1$	DLW$_2$	DLW$_3$	DLW$_4$	SCS$_5$
a/b	0.4	0.3	0.25	0.2	0.48
a/λ	0.163	0.118	0.097	0.077	0.10
r_{sh}, MΩ/m	24	30	34.35	41.4	28.5
α, m^{-1}	0.008	0.024	0.045	0.113	0.013
β_{gr}	0.051	0.018	0.009	0.0037	0.10

If we adopt the ratio of the disk aperture radius to the wavelength $a/\lambda \approx 0.1$, then an appropriate linac design with a large efficiency readily follows for given rf power and beam parameters.

Table 4.3 presents relevant data for such a DLW and an SCS for $P_0 = 10$ and 5 MW and $U = 5.8$ and 10 MV. Despite an advantage of the SCS in transverse dimensions(to be discussed later) the DLW option is more preferable in terms of longitudinal dimensions.

Now, we consider some aspects in choosing the geometry of a DLW for a 2×1 TeV e$^{\pm}$ linear collider [102]. We pose on the dependence of the DLW dynamic range and rf source parameters on the frequency and loading parameter a/λ.

The choice of a frequency of a linear collider [103, 104] is conditioned by the requirements to separate system components,

Table 4.3 Efficiency η and structure length L for a DLW and SCS at 915 MHz with $a/\lambda \approx 0.1$, $\theta = 2\pi/3$, and $\beta_{ph} = 1$.

P, MW	U, MV	DLW		SCS	
		η, %	L, m	η, %	L, m
	10	75	2.7	80	5.4
10	8	77	2.1	81	4.3
	5	80	1.3	83	2.7
	10	70	3.8		
5	8	73	3.0	79	6.1
	5	78	1.9	82	3.8

including the accelerating waveguide, rf sources, damping rings, tolerances for the manufacture and assembly, limitations on the luminosity and losses for synchrotron radiation in the collision, and some other parameters in the final focus.

In order to limit the number of independent variables we assume that some parameters are constant. First of all, we set the electric field strength at $E_0 = 100$ MV/m [102]. We assume also that the collider operates in the single bunch mode with a repetition rate of $f_r = 100$ Hz. We will consider sections of a DLW excited in the $2\pi/3$ mode with a constant impedance.

Table 4.4 presents the electrodynamic characteristics of a DLW section with constant impedance and excited in the $2\pi/3$ mode for four frequencies 10, 14, 18, and 30 GHz. The dimensions a and b of the DLW, group velocity β_{gr}, shunt impedance per unit length r_{sh}, attenuation parameter αL, overvoltage coefficient k_{ov}, and the rf filling time of a 1-m section, all these parameters have been calculated with the data of our handbook [3].

The pulse power per unit length has been calculated by the expression

$$\frac{P_p}{L} = \frac{E_0^2}{2 r_{sh} \alpha \eta_S L}, \tag{4.14}$$

where

$$\eta_S = \left(\frac{1 - e^{-\tau}}{\tau} \right)^2. \tag{4.15}$$

If the DLW is fed from an rf source through a compression systems, then

$$\frac{P_p}{L} = \frac{E_0^2}{2 r_{sh} \tau \eta_S C_f \eta_c}, \tag{4.16}$$

where η_c is the efficiency of pulse compression, and C_f is the ratio of the rf pulse lengths at the input and output from the compression circuit.

The average power of the rf source per unit length has been calculated by the formula

$$\frac{P_{av}}{L} = f_r t_f \frac{P_p}{L} F_f = \frac{E_0^2 f_r}{\eta_S \omega \frac{r_{sh}}{Q}} F_f \,, \qquad (4.17)$$

where $F_f = 1 - \beta_{gr}$.

Table 4.4 Electrodynamic characteristics of a 1-m constant-impedance section excited in the $2\pi/3$ mode with $E_0 = 100$ MV/m and $f_r = 100$ Hz for four frequencies 10, 14, 18, and 30 GHz.

	10 GHz	14 GHz	18 GHz	30 GHz
λ, cm	3.0	2.14	1.67	1.0
a/λ	0.17	0.17	0.17	0.17
	0.21	0.21	0.21	0.21
a, cm	0.51	0.36	0.28	0.17
	0.63	0.45	0.35	0.21
a/b	0.4123	0.4123	0.4123	0.4123
	0.4869	0.4869	0.4869	0.4869
β_{gr}	0.063	0.063	0.063	0.063
	0.109	0.109	0.109	0.109
r_{sh}, MΩ/m	75.1	89.04	100.7	130.0
	57.8	68.4	77.5	100.0
$\tau = \alpha L$, N	0.27	0.45	0.65	1.4
	0.135	0.225	0.325	0.7
k_{ov}	2.51	2.51	2.51	2.51
	2.8	2.8	2.8	2.8
t_f, ns	52.9	52.9	52.9	52.9
	30.59	30.59	30.59	30.59
η_s	0.768	0.648	0.54	0.289
	0.875	0.802	0.728	0.517
P_p/L, MW/m	332	200	148	97
	758	424	284	141
P_{av}/L, kW/m	1.646	0.891	0.732	0.480
	2.06	1.156	0.774	0.382

Table 4.5 Electrodynamic characteristics of a constant impedance DLW at the $2\pi/3$ mode, $f = 14$ GHz, $L = 1$ m, $f_r = 100$ Hz, $E_0 = 100$ MV/m for five iris apertures.

	a/b				
	0.2	0.3	0.411	0.4	0.5
a/λ	0.077	0.118	0.169	0.164	0.218
β_{gr}	0.005	0.021	0.062	0.057	0.119
α, m^{-1}	6.39	1.31	0.388	0.479	0.224
r_{sh}, MΩ/m	170.2	126.1	88.5	91.8	67.4
Q	6201	6183	6041	6013	5972
k_{ov}	1.79	1.96	2.64	2.62	3.00
t_f, ns	654	157	50	59	30
r_{sh}/Q, kΩ/m	27.5	20.4	14.7	15.3	11.3
E_{max}, MV/m	179	196	264	262	300
T_0, ns	141.0	140.6		136.7	135.8
η_s	0.02	0.3	0.69	0.63	0.80
P_p/L, MW/m ($E_0 = 100$ MV/m)	193	197.6	200	180.5	413.8
P_{av}/L, kW/m ($f_c = 100$ Hz)	1.73	1.77	1.12	1.12	1.12
E_0, MV/v	93	140.6	88.6	105.6	69.4
$\partial f/\partial 2b$, MHz/μm	−0.85	−0.86	−0.88	−0.88	−0.90
$\partial f/\partial 2a$, MHz/μm	0.1	0.2	0.36	0.34	0.48
$\partial f/\partial d$, MHz/μm	−0.01	−0.05	−0.15	−0.14	−0.25
$\partial f/\partial t$, MHz/μm	−0.010	−0.078	−0.189	−0.186	−0.314
∂q_i, μm for $\Delta U/U = 1\%$	0.55	2.21	6.1	5.40	9.51

As can be seen from Table 4.4, at higher frequencies a lower rf power is required to obtain the accelerating field 100 MV/m, and one can easily achieve the necessary gradient of the electric field because the overvoltage factor is proportional to $f^{1/3}$ [25].

However, higher frequencies imply a number of technological disadvantageous, e.g., tough tolerances in DLW manufacturing and assembly, and, what is especially unfortunate, they lead to transverse and longitudinal instabilities associated with the wake fields. It is known that the transverse wake fields, leading to "head–tail" instabilities, are proportional to f^3, while the longitudinal wake fields, broadening the energy spectrum, are proportional to f^2.

The effect of wake fields on the transverse and longitudinal beam stability can be reduced by increasing the parameter a/λ in single bunch designs. Such structures with an increased group velocity allow for a somewhat lower average rf power requirements, slacken the DLW dimensional tolerances, and improve the DLW vacuum and cooling performance.

Table 4.4 gives the characteristic parameters of a DLW and rf source for two values of a/λ, and Table 4.5 gives these parameters for a fixed frequency of 14 GHz and five values of a/λ thus covering all the range of characteristic parameters presented in our handbook [3]. In addition to parameters summarized in Table 4.4, this table gives the electric field strength 200 MV/m in the pulse mode, and dimensional tolerances for a bunch energy spread in a 1-m section within 1%.

An analysis of tabulated data indicates that a DLW design with $a/\lambda = 0.17$ operating at 14 GHz with an acceleration rate of 100 MV/m would require 200 MW of rf power with a pulse length of 50 ns, $P_{av} = 1$ kW, and $f_r = 100$ Hz.

4.2 Couplers for Linear Colliders

An important problem in the development of an accelerating structure of an electron–positron linear collider is the design of a matched coupler. This device must operate at the given rf frequency so that the electromagnetic field would be symmetric near the beamline in the coupler. Several designs satisfying these requirements are known. One of them uses a cutoff or short-circuited waveguide placed opposite to the feeding rectangular waveguide [105]. Another design includes two power feeders [106] placed on either side of the coupler cavity and a waveguide bridge to divide the rf power.

The coupler designed for the DESY linear collider project [107, 108] was configured as the first cell of a disk-loaded

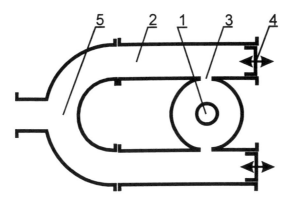

Figure 4.2 Coupler for the DESY linear collider project

waveguide(labeled *1* in Fig. 4.2). The rectangular waveguides (*2*) are parallel and have coupling apertures (*3*) in their narrow walls. Each rectangular waveguide is short circuited (*4*) at one of its end and is connected to a port of a T-junction (three-port power divider) at the other end (*5*).

To obtain expressions for the development of the matching method and calculation of impedance characteristics, we used the resonant model of a DLW and a coupler [108]. DLW cells are characterized by a resonance frequency f_r and the coupling coefficient with adjacent cell $k_0/2$. The coupler is characterized by the resonance frequency f_c, coupling coefficient with an adjacent cell $k_c/2$, coupling coefficient with a rectangular waveguide β, and quality factor Q_c. In the general case, $k_c/2 \neq k_0/2$. For the appropriately chosen T-junction reference planes, the T-junction may be characterized by the scattering matrix with elements $S_{11} = 0$, $S_{22} = S_{33} = 1/2$, $S_{12} = S_{21} = S_{13} = S_{31} = 2^{-1/2}$, and $S_{23} = S_{32} = -1/2$. The positions of the short-circuited plungers in the rectangular waveguides are characterized by $\psi = \pi - 4\pi\Delta l/\lambda_w$, where λ_w is the wavelength in the rectangular waveguide, Δl is the distance between the short-circuited plunger and the appropriately chosen reference plane so that it is negative if the plungers are placed outside the coupler and positive if they placed inside the coupler.

To secure the traveling wave regime in an infinite uniform lossless DLW with mode θ_0 at operating frequency f_0, it is necessary that

$$f_r = \frac{f_0}{\sqrt{1 - k_0 \cos \theta_0}} . \qquad (4.18)$$

For a coupler connected with an infinite uniform lossless DLW, the reflection coefficient at the input of the T-junction is defined by [108]

$$\Gamma = \frac{2\beta - \left(2\beta - 1 - \beta_{DLW} - jQ_c \left[\dfrac{f}{f_c} - \dfrac{f_c}{f} + \dfrac{f_c}{f}\dfrac{k_0}{2}\left(\dfrac{k_c}{k_0}\right)^2 \cos\theta\right]\right) \exp(j\psi)}{2\beta + 1 + \beta_{DLW} + jQ_c \left[\dfrac{f}{f_c} - \dfrac{f_c}{f} + \dfrac{f_c}{f}\dfrac{k_0}{2}\left(\dfrac{k_c}{k_0}\right)^2 \cos\theta\right] - 2\beta \exp(j\psi)}$$

$$(4.19)$$

where j is the imaginary unity and f and θ are related by the dispersion characteristic of an infinite uniform lossless DLW

$$f = f_r \sqrt{1 - k_0 \cos\theta} ,$$

$$\beta_{DLW} = \frac{k_0}{2}\left(\frac{k_c}{k_0}\right)^2 \frac{f_c}{f} \sin\theta . \qquad (4.20)$$

The reflection coefficient vanishes at f_0 if

$$f_c = \frac{f_0}{\sqrt{1 - \dfrac{k_0}{2}\left(\dfrac{k_c}{k_0}\right)^2 \cos\theta_0}} ,$$

$$\psi = \pi \,(\text{or } \Delta l = 0), \qquad (4.21)$$

$$4\beta = 1 + Q_c \frac{k_0}{k_c}\left(\frac{k_c}{k_0}\right)^2 \frac{\sin\theta_0}{\sqrt{1 - \frac{k_0}{2}\left(\frac{k_c}{k_0}\right)^2 \cos\theta_0}} \, .$$

Since $4\chi \gg 1$ (Q_c is usually large) we can assume that the coupler loaded quality factor is

$$Q_{cL} = \frac{Q_c}{4\beta} = \frac{2}{k_0}\left(\frac{k_0}{k_c}\right)^2 \frac{\sqrt{1 - \frac{k_0}{2}\left(\frac{k_c}{k_0}\right)^2 \cos\theta_0}}{\sin\theta_0} \tag{4.22}$$

Thus, to match the coupler, one must experimentally determine the position of the short-circuiting plunger in the rectangular waveguide at which $\theta = \pi$, as well as the coupler parameters f_c, $k_c/2$, (or k_c/k_0) and Q_{cL}. The DLW cell parameters (θ_0, f_0, k_0, f_r) are assumed to be known. If the first DLW cell (adjacent to the coupler) is strongly detuned (for example by inserting a thick ring into this cell), the reflection coefficient can be rewritten as

$$\Gamma = \frac{2\beta - \left(2\beta - jQ_c\left[\frac{f}{f_c} - \frac{f_c}{f}\right]\right)\exp(j\psi)}{2\beta + jQ_c\left[\frac{f}{f_c} - \frac{f_c}{f}\right] - 2\beta\exp(j\psi)} \tag{4.23}$$

and, if the coupler is strongly detuned by inserting a moving cylindrical plunger into the coupler,

$$\Gamma_0 = \exp(j\psi). \tag{4.24}$$

From (4.23) and (4.24) we have $\Gamma_1(f = f_c) = 1$ and $\Gamma_0(f = f_c) = \exp(j\psi)$. Moreover, if $\psi = \pi$, then

$$\Gamma_1 = (f = f_c \pm \Delta f) = \exp(\pm j\alpha),$$

where

$$\alpha = \arctan\left(\frac{Q_c}{2\beta}\frac{\Delta f}{f_c}\right)$$

and $\Delta f \ll f_c$.

Thus, by measuring the arguments of the reflection coefficients Γ_1 and Γ_0 one can experimentally determine the coupler resonance frequency f_c and the loaded quality factor $Q_c/(4\beta)$.

If the second DLW cell is strongly detuned by inserting a thick ring into this cell and the resonance frequency of the first DLW cell is made equal to f_c (for example by inserting a thin ring into the first cell), there are two frequencies f_1 and f_2 at which $\arg \Gamma_2(f = f_{1,2}) - \arg \Gamma_0(f = f_{1,2}) = \pm\pi$. Here, Γ_2 is the reflection coefficient for the strongly detuned second cell. The coupling coefficient between the coupler and the first DLW cell can be determined as

$$\frac{k_c}{2} = \frac{|f_1^2 - f_2^2|}{f_1^2 + f_2^2} \tag{4.25}$$

Notice that if the first-cell frequency were f_c we had $\arg \Gamma_2(f = f_c) = \arg \Gamma_0(f = f_c)$.

Expression (4.19) represents the impedance characteristic of the infinite uniform lossless section with an output coupler. If the section is not uniform and consists of N cells (including input and output couplers) the impedance characteristic can be written as

$$\Gamma = \frac{2\beta - \left(2\beta - 1 - jQ_c \frac{f_c}{f}\left[\frac{f^2}{f_c^2} - 1 + \frac{k_c}{2}\frac{x_1}{x_2}\right]\right)\exp(j\psi)}{2\beta + 1 + jQ_c \frac{f_c}{f}\left[\frac{f^2}{f_c^2} - 1 + \frac{k_c}{2}\frac{x_1}{x_2}\right] - 2\beta\exp(j\psi)}, \tag{4.26}$$

where $x_1, x_2, x_2, ..., x_N$, are complex variables, which can be calculated as

$$x_{n-1} = -\frac{k_n}{k_{n-1}}x_{n+1} - \frac{2}{k_{n-1}}\left(\frac{f^2}{f_n^2}-1-j\frac{f}{f_n}\frac{1}{Q_n}\right)x_n. \qquad (4.27)$$

Here $k_n/2$ is the coupling coefficient between the nth and $(n-1)$th cells, f_n, Q_n are the resonance frequency and unloaded quality factor of the nth cell, Q_1 is the loaded quality factor of the output coupler cell,

$$f_1 = f_c, \quad Q_1 = Q_c, \quad \frac{k_1}{2} = \frac{k_c}{2},$$

$$x_N = 1, \quad k_N = 0, \quad x_{N+1} = 0, \quad n = N, N-1, N-2, ..., 2.$$

The physical meaning of the quantity $x_n = |x_n|\exp(j\theta_n)$ is as follows: $|x_n| = \sqrt{2W_n}$, where W_n is the electrical field energy stored in nth cell, and θ_n is the electrical field phase in the middle of the cell.

The reflection coefficient vanishes at the operational frequency f_0 if $\psi = \pi$ (or $\Delta l = 0$),

$$f_1 = f_c = \frac{f_0}{\sqrt{1-\frac{k_c}{2}\mathrm{Re}\left(\frac{x_2}{x_1}\right)}}, \qquad (4.28)$$

$$Q_{cL} = \frac{Q_c}{4\beta} = -\frac{2}{k_c}\frac{\sqrt{1-\frac{k_c}{2}\mathrm{Re}\left(\frac{x_2}{x_1}\right)}}{\mathrm{Im}\left(\frac{x_2}{x_1}\right)},$$

where $\mathrm{Re}(x_2/x_1)$ and $\mathrm{Im}(x_2/x_1)$ are the real and imaginary parts of (x_2/x_1), which is calculated at the operational frequency. If $\psi = 0$ or $\psi = 2\pi$, then $\Gamma = 1$ and the coupler can not be matched. When $\psi = \pi$, the coupler matching aperture is minimal. If $0 < \psi < \pi$, the coupler can be matched, but the coupling matching aperture must be enlarged.

Table 4.6. Experimental and calculated electrodynamics characteristics of input and output couplers.

Coupler		k_c	f_c, MHz	Q_{cL}	k_0	f_{r1}, MHz	$\lvert\Gamma\rvert$
Input	cal		2981.19	51.3	.04525	2964.65	0.0
Input	exp	.0455	2981.05	52.5		2963.63	0.047
Output	cal		2992.68	170.4	.01423	2987.39	0.0
Output	exp	.0139	2992.70	173.5		2987.40	0.035

The theory of coupler matching and the method of experimental determination of the coupler parameters were tested at the DESY linear collider project [107]. For these purpose two DLW section with 11 cells were manufactured. In the first section, the cells were similar to the initial cell of the collider section with a DLW period of 33.33 mm, $a/\lambda = 0.10885$, $t/\lambda = 0.05$, and $k_0 = 0.01423$, where a and t are the disk radius and thickness, respectively. The second section was composed of cells which were identical to the last cell ($a/\lambda = 0.155$, $k_0 = 0.04525$). The operational frequency was $f_0 = 2.998$ MHz, and the operational mode was $\theta = 2\pi/3$. The experimentally determined and calculated parameters of the input and output cells are summarized in Table 4.6 [108]. The experimental parameters are related to the matched couplers. The frequency f_{r1} is the resonance frequency of the cell adjacent to the coupler. The experimental values of the reflection coefficient Γ were obtained by the method of movable absorbing load [3]. Matching was achieved by changing the inner coupler diameter $2b_c$ and the coupler aperture width $2a_c$.

4.3 Equivalent Scheme in the Dipole Mode.

The equivalent circuit of a DLW with a constant gradient in the dipole mode is shown in Fig 4.3 [109].

In this circuit, the series circuit (L_{1n}, C_{1n}, r_{1n}) represents the TM$_{11}$ electromagnetic field. On the one hand, the coupling between cells is realized by the magnetic field represented by the

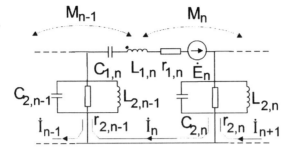

Figure 4.3 Equivalent circuit of a DLW cell at dipole mode.

mutual inductance M_n. On the other hand, there is a resonance type of coupling by the TE$_{11}$ electromagnetic field that is represented by the parallel circuit (L_{2n}, C_{2n}, r_{2n}). The cell excitation is modeled by a complex e.m.f. E_n in the series branch. Here, I_n are the complex loop currents.

In what follows, it would be more convenient to operate with parameters expressed in terms of L, C, and r as follows

$$f_{1n} = \frac{1}{2\pi\sqrt{L_{1,n}C_{1n}}}; \quad f_{2n} = \frac{1}{2\pi\sqrt{L_{2,n}C_{2n}}};$$

$$\frac{K_n}{2} = \frac{M_n}{\sqrt{L_{1,n}L_{1,n+1}}}; \quad \frac{K_{12n}}{2} = \frac{\sqrt{C_{1,n}C_{1,n+1}}}{C_{2,n}};$$

$$Q_{1n} = \sqrt{\frac{L_{1,n}}{C_{1,n}}}\frac{1}{r_{1,n}}; \quad Q_{2n} = \sqrt{\frac{C_{2,n}}{L_{2,n}}}r_{2n};$$

$$X_n = \sqrt{L_{1,n}}I_n;$$

(4.29)

By using these parameters we can obtain the system of equations in the complex variables X_n

$$\begin{bmatrix} A_1 & B_1 & 0 & . & . & & 0 \\ B_1 & A_2 & B_2 & . & . & & 0 \\ . & . & . & . & . & . & . \\ 0 & . & . & . & B_{N-2} & A_{N-1} & B_{N-1} \\ 0 & . & . & . & 0 & B_{N-1} & A_N \end{bmatrix} \begin{bmatrix} X_1 \\ X_2 \\ \vdots \\ X_{N-1} \\ X_N \end{bmatrix} = \begin{bmatrix} \dfrac{E_1}{j\omega\sqrt{L_{11}}} \\ \vdots \\ \vdots \\ \vdots \\ \dfrac{E_N}{j\omega\sqrt{L_{1N}}} \end{bmatrix}, \quad (4.30)$$

where $\omega = 2\pi f$ is the section excitation frequency,

$$A_n = 1 - \frac{f_{1n}^2}{f^2} - j\frac{f_{1n}}{f}\frac{1}{Q_{1n}} - \frac{f_{1n}^2}{f^2}\left[\sqrt{\frac{C_{1n}}{C_{1,n-1}}}\frac{K_{12,n-1}}{2}\frac{1}{1 - \frac{f_{2,n-1}^2}{f^2} - j\frac{f_{2,n-1}}{f}\frac{1}{Q_{2,n-1}}}\right.$$

$$\left. + \sqrt{\frac{C_{1,n}}{C_{1,n+1}}}\frac{K_{12,n}}{2}\frac{1}{1 - \frac{f_{2,n}^2}{f^2} - j\frac{f_{2,n}}{f}\frac{1}{Q_{2,n}}}\right], \quad (4.31)$$

$$B_n = \frac{K_n}{2} + \frac{K_{12,n}}{2}\frac{f_{1,n}f_{1,n+1}}{f^2}\frac{1}{1 - \frac{f_{2,n}^2}{f^2} - j\frac{f_{2,n}}{f}\frac{1}{Q_{2,n}}}. \quad (4.32)$$

At $n = 1$, the multiplier $\sqrt{C_{1,n}/C_{1,n-1}}\,K_{12,n-1}/2$ in A_1 must be replaced with $K_{12,0}/2 = C_{1,1}/C_{2,0}$, whereas at $n = N$, the multiplier $\sqrt{C_{1,N}/C_{1,N+1}}\,K_{12,N}/2$ must be replaced with $K_{12,N}/2 = C_{1,N}/C_{2,N}$.

Note that, in the equivalent circuit, the number of parallel circuits is equal to $N + 1$, whereas that of series circuits equals N, where N is the number of cells in the section. They are introduced for modeling the beam pipe in the input and output couplers as the first and the last parallel circuits (indexes 0 and N).

Voltages across the capacitive element C_{1n} ($U_{z,n}$) and C_{2n} ($U_{r,n}$) are related to X_n by the following expressions :

$$U_{z,n} = -j\frac{f_{1,n}}{f}\frac{X_n}{\sqrt{C_{1,n}}}, \quad (n = 1,2,...,N) \tag{4.33}$$

$$U_{r,n} = \frac{K_{12,n}}{2}\frac{1}{1-\frac{f_{2,n}^2}{f^2}-j\frac{f_{2,n}}{f}\frac{1}{Q_{2,n}}}\left(\sqrt{\frac{C_{1,n}}{C_{1,n+1}}}U_{z,n} - \sqrt{\frac{C_{1,n+1}}{C_{1,n}}}U_{z,n+1}\right),$$

$$(n = 1,2,...,N-1)$$

$$\tag{4.34}$$

$$U_{r,0} = -\frac{K_{12,0}}{2}\frac{1}{1-\frac{f_{2,0}^2}{f^2}-j\frac{f_{2,0}}{f}\frac{1}{Q_{2,0}}}U_{z,1} \tag{4.35}$$

$$U_{r,N} = \frac{K_{12,N}}{2}\frac{1}{1-\frac{f_{2,N}^2}{f^2}-j\frac{f_{2,N}}{f}\frac{1}{Q_{2,N}}}U_{z,N} \tag{4.36}$$

Using this equivalent circuit, we obtain the dispersion relation for an infinite uniform lossless DLW ($f_{1,n} = f_1$, $f_{2,n} = f_2$, $K_n = K$, $K_{12,n} = K_{12}$, $Q_{1,n} = \infty$, $Q_{2,n} = \infty$) in the form

$$f = \sqrt{\frac{F(\varphi)\pm\sqrt{F^2(\varphi)-4f_1^2f_2^2(1+K\cos\varphi)}}{2(1+K\cos\varphi)}}, \tag{4.37}$$

where $F(\theta) = f_1^2[1+K_{12}(1-\cos\theta)]+f_2^2(1+K\cos\theta)$, θ is the phase shift per DLW cell, and f is the frequency corresponding to θ.

The upper sign in (4.37) corresponds to the higher branch of the dispersion relation whereas the lower sign corresponds to the lower branch.

Denoting by f_{01} and $f_{\pi 1}$ the frequencies of the 0 and π modes of the lower branch and by f_{02} and $f_{\pi 2}$ the similar frequencies for the upper branch of the dispersion curve, we can write the expressions for f_1, f_2, K, and K_{12} as follows:

$$f_1 = f_{01}\sqrt{1+K}, \quad f_2 = f_{02}, \qquad (4.38)$$

or

$$f_1 = f_{02}\sqrt{1+K}, \quad f_2 = f_{01}, \qquad (4.39)$$

$$K = \frac{(f_{\pi 1}f_{\pi 2})^2 - (f_{01}f_{02})^2}{(f_{\pi 1}f_{\pi 2})^2 + (f_{01}f_{02})^2} = \frac{2M}{L_1}, \qquad (4.40)$$

$$K_{12} = \frac{1}{2}\left[(1-K)\frac{f_{\pi 1}^2 + f_{\pi 2}^2 - f_2^2}{f_1^2} - 1\right] = \frac{2C_1}{C_2}. \qquad (4.41)$$

The input coupler was modeled by adding a coupler cell. The frequency $f_{1,c}$ and quality factor $Q_{1,c}$ of this additional cell were chosen so as to obtain the traveling wave mode in the DLW consisting of cells similar to the first one. The frequency of the traveling wave mode was f_{op}, and the phase shift per cell at this frequency was θ_{0p}.

Using equations (4.30), (4.32), and (4.33) we calculated U_z as a function of the cell number at different frequencies for an SBLC accelerating section with constant gradient [107] consisting of first 30 cells (see Fig. 4.4). Curves 1 at these graphs correspond to the case when the input coupler is mismatched with first section cell . Curves 2 correspond to the case when the input coupler is matched with the DLW at the frequency $f_{0p} = 4.151$ MHz ($\theta_{0p} = 95°$). Excitation of the accelerating section was simulated by introducing an e.m.f. into cells 2, 8, 13 and 18 at 4.13760 GHz, 4.14495 GHz, 4.15096 GHz, and 4.15908 GHz, respectively. At these frequencies similar patterns were obtained using small perturbations in the first 30 cells for the mismatched input coupler [110]. In all calculations, the resonant frequency f_{11} of the series circuit was chosen at 4.84678 GHz and its quality factor Q_{11}

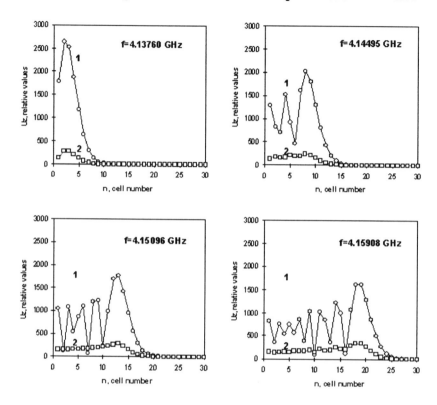

Figure 4.4. U_z versus cell number for the first 30 cells of an SBLC DLW at different frequencies. Curves *1* correspond to the mismatched input coupler; curves *2* correspond to the matched input coupler.

was 14.677. These values correspond to the lowest resonant frequency of the coupler cell $f_c = 4.15222$ GHz and its quality factor $Q_c = 159.77$.

Figure 4.4 demonstrates that matching the coupler in the hybrid mode reduces the amplitude of the electric field by more than five times in a wide frequency range.

4.4 Biperiodic Structures

According to the classification of biperiodic structures (BPS) outlined in Chapter 1, all BPSs, except disk and washer structures (DAW), differ in the design of coupling cells and coupling elements between adjacent accelerating cells. The profile of accelerating cells is optimized with respect to their shunt impedance and quality factor. The electrodynamic characteristics of accelerating structures, whose profiles are depicted in Figs. 4.5 and 3.3, are summarized in our handbook [4] as functions of their dimensions.

A sectional view of a BPS with on-axis cylindrical coupling cells and optimized accelerating cells is depicted in Fig. 3.3. The relative cell dimensions q_i/λ_0 form the functional [4]

$$\frac{a}{\lambda_{0a}} = F\left(\frac{R_a}{\lambda_{0a}}, \frac{a_1}{\lambda_{0a}}, \frac{L_a}{\lambda_{0a}}, \frac{d_g}{\lambda_{0a}}, \frac{l_{l1}}{\lambda_{0a}}, \frac{l_{l2}}{\lambda_{0a}}, \frac{R_r}{a_1 - a}, \eta, \chi, k_c\right) \quad (4.42)$$

for accelerating cells and

$$\frac{a}{\lambda_{0c}} = F\left(\frac{R_c}{\lambda_{0c}}, \frac{L_c}{\lambda_{0c}}, k_c\right) \quad (4.43)$$

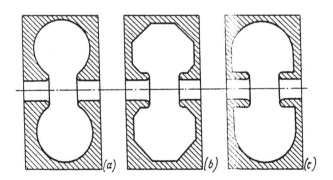

Figure 4.5. Profiles of accelerating cells of BPSs

for coupling cells. Here, λ_{0a} and λ_{0c} are the wavelengths of the eigenoscillation TM_{010} mode for accelerating and coupling cells, L_a and L_c are the lengths of accelerating and coupling cells, a is the radius of the beam channel, R_a and R_c are the radii of the accelerating and coupling cells, a_1 is the radius of a drift tube, d_g is the length of the accelerating gap, R_r is the radius of rounded corners of drift tubes, χ and η are the profiling angles, l_{l1} is a leg at angle χ, l_{l2} is a leg at angle η, $D = L_a + L_c + 2t$ for an on-axis coupling cell structure , and $D = L_a + t$ for an external coupling cell structure. From (3.19) we have for the coupling coefficient

$$k_c = F\left(\frac{r_{sl}}{R_{a,c}}, \frac{l_{sl}}{\lambda_0}, \frac{\Delta}{\lambda_0}, \frac{t}{\lambda_0}\right), \tag{4.44}$$

where r_{sl} is the radius of the symmetric axis of azimuthal coupling slot, l_{sl} is the length of a coupling slot, t is the thickness of the wall in which the slot is cut, λ_0 is the resonance wavelength in an accelerating cell or coupling cell, and Δ is the width of a coupling slot.

We investigated the following BPS characteristics, written with allowance for the frequency dependence,

$$\frac{E_z(0)\lambda_{1/2}^{3/4}}{\sqrt{P}} = F\left(\frac{q_i}{\lambda_{1/2}}\right),$$

$$\lambda_{1/2}^{1/2} r_{sh,eff} = F\left(\frac{q_i}{\lambda_{1/2}}\right),$$

$$\lambda_{1/2}^{-1/2} Q = F\left(\frac{q_i}{\lambda_{1/2}}\right),$$

$$\frac{1}{f_{1/2}^2}\frac{\partial f_{1/2}}{\partial q_i} = F\left(\frac{q_i}{\lambda_{1/2}}\right),$$

$$k_{\mathrm{p}} = F\left(\frac{q_i}{\lambda_{1/2}}\right).$$

These expressions are independent of $\beta_{\mathrm{ph}1/2}$ since, in the $\pi/2$ mode, $\beta_{\mathrm{ph}1/2} = 2D/\lambda_0$.

Obviously, it is almost impossible to tabulate the BPS characteristics experimentally in the entire feasible range of relative dimensions.

As a way out, for collecting BPS reference data, we fixed parameters that weakly influence the respective functional and tabulated the above relations for two (at most three) independent variables which contribute more than others in these functionals. The secondary parameters must be fixed at a level which provides the extreme effective shunt impedance of the accelerating system. This problem has been solved with the use of computer codes GNOM [62] and MULTIMODE [77] for axially symmetric cavities.

In order to determine the frequency of a mode in a chain of coupled cavities terminated with half-length accelerating cells, we resort to the expression [7]

$$k_{\mathrm{c}}^2 \cos^2\theta = \left(1 - \frac{f_{\mathrm{a}}^2}{f_n^2} + k_{\mathrm{c,a}}\cos 2\theta\right)\left(1 - \frac{f_{\mathrm{c}}^2}{f_n^2} + k_{\mathrm{c,c}}\cos 2\theta\right). \qquad (4.45)$$

Here, f_{a} and f_{c} are the resonant frequencies of accelerating and coupling cells, respectively; $k_{\mathrm{c,a}}$ and $k_{\mathrm{c,c}}$ are the second-nearest-neighbor coupling coefficient between adjacent accelerating cells and nearest coupling cells, respectively; $\theta = \pi n/(N-1)$ is the mode; N is the full number of cells in the chain; and $n = 0, 1, 2, \ldots, N-1$.

If $f_{\mathrm{a}} = f_{\mathrm{c}} = f_0$ and $k_{\mathrm{c,a}} = k_{\mathrm{c,c}} = 0$, then equation (4.45) can be rewritten as

$$f_n = f_0\left(1 + k_{\mathrm{c}}\cos\theta\right)^{-1/2}, \qquad (4.46)$$

where f_0 is the resonant frequency of cells with allowance for the effect of coupling slots.

We now illustrate how the BPS reference data can be used in dimensioning an accelerating cavity for a high-current linac.

For a SW linac, the energy increment (in volts) in a section of length L with a shunt impedance per unit length $r_{sh,eff}$ can be determined by the formula [88]

$$U = \sqrt{r_{sh,eff} L P_{rf}} \, \frac{2\sqrt{\beta_0}}{1+\beta_0} - \frac{I_0 r_{sh,eff} L}{1+\beta_0}, \tag{4.47}$$

where β_0 is the initial coupling coefficient of the cavity with a waveguide (in the absence of beam load), and P_{rf} is the rf generator power. This expression is written for the frequency compensation regime of the beam reactance when the bunch is in the optimal accelerating phase. The efficiency for such a linac can be calculated as

$$\eta = \frac{2\sqrt{\beta_0}}{1+\beta_0} \sqrt{\frac{r_{sh,eff} L}{P_{rf}}} I_0 - \frac{r_{sh,eff} L}{P_{rf}(1+\beta_0)} I_0^2. \tag{4.48}$$

Clearly, for given values of P_{rf}, and $R_{sh,eff} = r_{sh,eff} L$, the dependence on η and I_0 is a parabola with parameter β_0.

If we fix I_0 and take the derivative $\partial \eta / \partial \beta_0$, then the optimum coupling coefficient will be

$$\beta_{opt} = \left(\frac{I_0}{2} \sqrt{\frac{r_{sh,eff} L}{P_{rf}}} + \sqrt{1 + \frac{I_0^2}{4} \frac{r_{sh,eff} L}{P_{rf}}} \right)^2. \tag{4.49}$$

In the operating regime, a structure with the optimal coupling coefficient β_{opt} operates with the critical coupling. Accordingly, equations (4.47) and (4.48) can be rewritten as

$$U = \frac{\sqrt{r_{sh,eff} L P_{rf}}}{\dfrac{I_0}{2} \sqrt{\dfrac{r_{sh,eff} L}{P_{rf}}} + \sqrt{1 + \dfrac{I_0^2}{4} \dfrac{r_{sh,eff} L}{P_{rf}}}}, \tag{4.50}$$

$$\eta = \frac{I_0 \sqrt{\dfrac{r_{sh,eff}L}{P_{rf}}}}{\dfrac{I_0}{2}\sqrt{\dfrac{r_{sh,eff}L}{P_{rf}}} + \sqrt{1 + \dfrac{I_0^2}{4}\dfrac{r_{sh,eff}L}{P_{rf}}}}. \tag{4.51}$$

Figure 4.6 shows η as a function of $R_{sh,eff} = r_{sh,eff}L$ for different values of the coupling coefficient in a linac with $P_{rf} = 10$ MW and $U = 10$ MV. Choosing a large initial coupling coefficient, one naturally obtains a large efficiency; however, it would be unreasonable to take it larger than 2.5 in view of the difficulties associated with the transient regime [111].

For the case of $\beta_{opt} = 2.5$, Fig. 4.7 shows η versus $R_{sh,eff}$ for six configurations of a linac rated at 915 MHz. For a given $r_{sh,eff}$ and rf generator power, these plots can be used to estimate the length of a cavity yielding the desired energy and maximum efficiency.

Table 4.7 presents the values of η and L calculated by formulas (4.50) and (4.51) for an SW linac with biperiodic structure characterized by $r_{sh,eff} = 45$ MΩ/m at $\beta_0 = 2.5$. It would be interesting to compare these data with a TW linac calculated for the same rf power and beam parameters (Table 4.3). An obvious

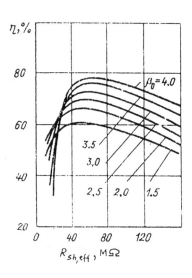

Figure 4.6. Efficiency of a SW linac with a BPS against the shunt impedance for different values of rf input coupling.

Figure 4.7. Efficiency of a standing wave BPS linac as a function of the effective shunt impedance plotted for different increments of energy at $P_{rf} = 5$ MW (solid curves) and 10 MW (dashed curves).

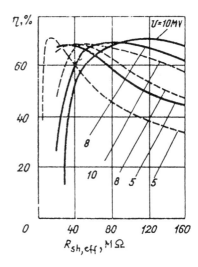

advantage of the standing wave regime in biperiodic structures is an approximately halved length of the accelerating structure.

At a frequency of 915 MHz, one may realize the shunt impedance $r_{sh,eff} = 45$ MΩ/m in different BPSs. Table 4.8 summarizes the characteristics of BPSs with coaxial coupling cells for two configurations: (1) shown in Fig. 1.4d and (2) a design with a reduced radius. A more detailed consideration of these characteristics will be given below. The values of shunt impedance

Table 4.7 Values of η and L for a linac with biperiodic structure characterized by $r_{sh,eff} = 45$ MΩ/m at 915 MHz and $\beta_0 = 2.5$. $\pi/2$ mode, $\beta_{ph} = 1$.

P_{rf}, MW	U, MV	η, %	L, m
	10	72	0.9–1.2
5	8	72	0.65
	5	70	0.45
	10	72	1.6–2.5
10	8	72	1.2–1.5
	5	72	0.65

Table 4.8 BPS characteristics at 915 MHz and $\beta_{ph} = 1$.

	Coupling cell type		
	side	on-axis	ring
$r_{sh,eff}$	48 MΩ/m	45 MΩ/m	45 MΩ/m
k_c	5%	11%	18%
$k_c / k_{c,a}$	30	200	
Outer radius at 1.35 GHz	16.5 cm	9 cm	14.7 cm
Cooling technology	sophistic.	simple	moderate
Vacuum conductance	good	low	good
Beam excitation	in acceler. cells	acceler. & coupl. cells	in acceler. cells

	Coaxial coupling cells		DAW
	option 1	option 2	structure
$r_{sh,eff}$	45 MΩ/m	45 MΩ/m	74 MΩ/m
k_c, %	12%	12%	50%
$k_c / k_{c,a}$		110	10
Outer radius at 1.35 GHz	12.9 cm	12.4 cm	16.8 cm
Cooling technology	simple	simple	sophistic.
Vacuum conductance	low	low	excellent
Beam excitation	in acceler. cells	in acceler. cells	

per unit length at $a/\lambda_{1/2} = 0.05$ are summarized in Table 4.8, where $k_{c,a}$ is the second nearest-neighbor-coupling coefficient between accelerating cells.

For all BPSs except disk and washer structures, at 915 MHz and the coupling coefficient 5–15%, $r_{sh,eff}$ is within 40 to 50 MΩ/m [113]. In DAW structures, the coupling coefficient reaches 50%, however, its wide use in the development of high-current linacs is limited by the existence of HOMs near the TM_{020} operating mode, which can degrade the transverse beam stability at high accelerating currents.

Data presented in this section lead us to assert that the standing wave regime may be quite competitive with the traveling wave regime in the development of high-current linacs.

In this section, we have not considered such factors as the stability of rf power supply and the thermal regime in dissipation of a high average power, but they can be an obstacle in the development of a high-current SW linac based on biperiodic structures.

4.5 Rectangular Biperiodic Structures

Since the focusing of a relativistic electron beam by an axially symmetric SW accelerating structure is absent, this system is impractical for some applications. The reason is that the transverse momentum gained by a relativistic particle from the radial electric field at beam apertures is nearly offset by that from the magnetic field in the central region of the cavity. By braking the axial symmetry of the accelerating structure it is possible to change the balance of the transverse momentum, for example, so that in one plane it would gain a greater momentum from the electric field. In the other plane, the accelerating structure with quadrupole symmetry provides identical in magnitude, but opposite excess of the transverse momentum due to the magnetic field. Thus, accelerating structures with quadrupole symmetry will provide a phase dependent quadrupole-like focusing, which is rather strong and therefore promising for practical applications.

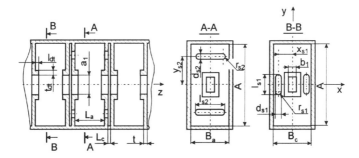

Figure 4.8. RBPS with rectangular beam apertures.

Rectangular (prismatic) biperiodic structures (RBPS) with circular [114] and rectangular [115] beam apertures (Fig. 4.8) offer both high accelerating gradients and strong transverse focusing. RBPSs considered here belong to the class of accelerating structures with quadrupole symmetry, and thus possesses quadrupole focusing properties. RBPSs possess the well known advantages of biperiodic accelerating structures with cylindrical cavities [116] and those of lesser known, periodic, axially-asymmetric structures with rectangular cavities and circular beam channels [2] and with circular cavities and rectangular beam channels [117, 118]. Specifically, RBPSs have a high effective shunt impedance, good frequency stability, and rf focusing properties.

The coupling coefficient and resonance frequency of this structure can be calculated using the equivalent circuit, shown at Fig. 4.9 for a rectangular $(A \times B)$ cavity excited in the TM_{110} mode.

Neglecting the accelerating half-cell coupling yields

$$\begin{cases} \left(1 - \dfrac{\omega_{01}^2}{\omega^2}\right)X_1 + \dfrac{k_c}{\sqrt{2}}X_2 = 0 \\[3mm] \dfrac{k_c}{\sqrt{2}}X_1 + \left(1 - \dfrac{\omega_{02}^2}{\omega^2}\right)X_2 + \dfrac{k_c}{\sqrt{2}}X_3 = 0 \end{cases} \qquad (4.52)$$

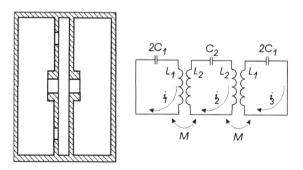

Figure 4.9. Resonator and its equivalent circuit.

$$\left| \frac{k_c}{\sqrt{2}} X_2 + \left(1 - \frac{\omega_{01}^2}{\omega^2}\right) X_3 = 0 \right.$$

where $k_c = M/\sqrt{L_1 L_2}$ is the coupling coefficient of an infinite structure and $X_1 = \sqrt{L_1} I_1$, $X_2 = \sqrt{L_2} I_2$, $X_3 = \sqrt{L_3} I_3$ are the normalized amplitudes of the field,

$$\omega_{01} = \frac{1}{\sqrt{2L_1 C_1}}, \quad \omega_{02} = \frac{1}{\sqrt{2L_2 C_2}}, \qquad (4.53)$$

L_1 and $2C_1$ are the half-cell inductance and capacitance, and L_2 and C_2 are the inductance and capacitance of a coupling cell.

Putting the determinant of equations (4.52) to zero

$$\begin{vmatrix} 1 - \dfrac{\omega_{01}^2}{\omega^2} & \dfrac{k_c}{\sqrt{2}} & 0 \\[3mm] \dfrac{k_c}{\sqrt{2}} & 1 - \dfrac{\omega_{02}^2}{\omega^2} & \dfrac{k_c}{\sqrt{2}} \\[3mm] 0 & \dfrac{k_c}{\sqrt{2}} & 1 - \dfrac{\omega_{01}^2}{\omega^2} \end{vmatrix} = 0,$$

$$\left(1 - \frac{\omega_{01}^2}{\omega^2}\right)\left[\left(1 - \frac{\omega_{02}^2}{\omega^2}\right)\left(1 - \frac{\omega_{01}^2}{\omega^2}\right) - \frac{k_c^2}{2}\right] - \frac{k_c}{\sqrt{2}}\left[\frac{k_c^2}{2}\left(1 - \frac{\omega_{01}^2}{\omega^2}\right)\right] = 0$$

$$(4.54)$$

yields the resonant frequencies

$$\omega_1 = \omega_{01}, \qquad (4.55)$$

$$\omega_{2,3}^2 = \frac{\omega_{01}^2 + \omega_{02}^2}{2(1 - k_c^2)} \pm \frac{1}{2(1 - k_c^2)}\sqrt{\left(\omega_{01}^2 - \omega_{02}^2\right)^2 + 4k^2 \omega_{01}^2 \omega_{02}^2}, \qquad (4.56)$$

so that the tuning criterion is

$$|\omega_2 - \omega_{01}| \approx |\omega_3 - \omega_{01}|.$$ (4.57)

Thus, ω_{01} is just the cavity's operating frequency undistorted by dispersion near the $\pi/2$ mode. From equation (4.56), assuming $\omega_{01} = \omega_{02} = \omega_0$, we have

$$\omega_{2,3}^2 = \frac{\omega_0^2(1 \pm k_c)}{1 - k_c^2}.$$ (4.58)

However, $k_c \ll 1$, therefore

$$\frac{\omega_2 - \omega_3}{\omega_0} \cong k_c.$$ (4.59)

Our prototype cavity can be exited at frequencies of 0, $\pi/2$, and π modes.

The coupling parameters for slots with dimensions of Fig. 4.8 and $2D = L_a = L_c + 2t$ are [115]

$$\left\{ \begin{array}{l} k_{c1} \cong \left(\dfrac{8Df^2}{c^2}\right)^2 \dfrac{Z_{sl}}{Z_0} \dfrac{\tilde{l}_{s1}^3}{3} \dfrac{1}{\sqrt{L_a L_c}} \left(\dfrac{A}{B}\right)^3 \left[1 + \left(\dfrac{A}{B}\right)^2\right]^{-2} \sin^2\left(\dfrac{\pi x_{s1}}{B}\right) \\[4mm] k_{c2} \cong \left(\dfrac{8Df^2}{c^2}\right)^2 \dfrac{Z_{sl}}{Z_0} \dfrac{\tilde{l}_{s2}^3}{3} \dfrac{1}{\sqrt{L_a L_c}} \dfrac{A}{B} \left[1 + \left(\dfrac{A}{B}\right)^2\right]^{-2} \sin^2\left(\dfrac{\pi y_{s2}}{A}\right) \end{array} \right.,$$ (4.60)

where the normalized characteristic impedance of a symmetric strip line, Z_{sl}/Z_0, may be calculated with equation (3.11).

Both pairs of slots must have the same coupling coefficients, therefore the discrepancy is

$$\frac{k_{c1}}{k_{c2}} \cong \left[\frac{\tilde{l}_{s1}}{\tilde{l}_{s2}}\right]^3 \left(\frac{A}{B}\right)^2 \left(\frac{\sin\left(\dfrac{\pi x_{s1}}{B}\right)}{\sin\left(\dfrac{\pi y_{s2}}{A}\right)}\right)^2 = 1.$$ (4.61)

The cell resonant frequency shifts induced by a coupling slot may be calculated from (3.14).

The cavity focusing is characterized by a gradient which can be calculated from the electric field measured by the bead-pull technique near the cavity axis or obtained with a 3-D rf cavity computer code. The normalized transverse focusing gradients are

$$
G_x(x,0)_n = \frac{\partial^2 E_{z1}(x,0)/E_{z1}(0,0)}{\partial x^2},
$$

$$
G_y(0,y)_n = \frac{\partial^2 E_{z1}(0,y)/E_{z1}(0,0)}{\partial y^2},
$$

(4.62)

and the transverse focusing gradients are

$$
G_x(x,0) = \frac{1}{\omega} G_x(x,0)_n E_{z1}(0,0)\sin\varphi
$$

$$
G_y(0,y) = \frac{1}{\omega} G_x(0,y)_n E_{z1}(0,0)\sin\varphi
$$

(4.63)

where φ is the bunch phase with respect to the rf field such that at $\varphi = 0$ the energy of a relativistic particle is a maximum, $E_{z1}(x,y)$ is the first Fourier spatial harmonic of the longitudinal electric field $E_z(x,y,z)$.

Figure 4.10 shows the normalized horizontal (G_x) and vertical (G_y) focusing gradients, and the effective shunt impedance for an RBPS with rectangular beam apertures, but without coupling slots. The electrodynamics characteristics depend on the ratios a_1/b_1 and A/B defined in Fig. 4.8. For the $\pi/2$ mode and the phase velocity equal to the speed of light ($\beta_{ph} = 1$) at 2450 MHz, we have obtained $-G_x$ and G_y as high as ~1000/m^2 with a large A/B and a square or circular ($a_1/b_1 = 1$) beam hole by vertically elongating the structure. For $A/B \approx 1$ and a large a_1/b_1, the focusing gradients have identical magnitudes but opposite signs.

It is known that an increase of a intercell coupling of a biperiodic structure decreases its shunt impedance since it changes the surface current distribution in both coupling and

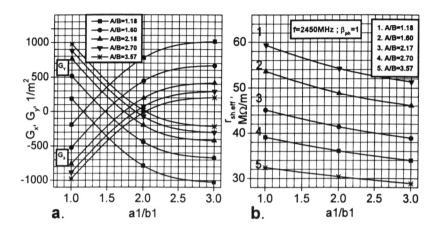

Figure 4.10. Focusing gradients and shunt impedance versus a_1/b_1.

accelerating cells. This behavior can be seen in Fig. 4.11. However, by shifting coupling slots to the axis, one may obtain a coupling factor as high as 15% without degrading the shunt impedance. We have investigated three highly coupled rectangular structures whose dimensions are given in Table 4.9. Design (a) had rectangular beam apertures and coupling slots,

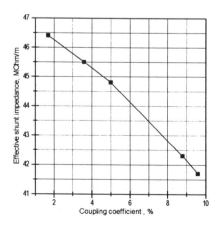

Figure 4.11. Effective shunt impedance versus the coupling coefficient for an RBPS with rectangular beam channel (2450 MHz, $\beta_{ph} = 1$)

Table 4.9. Structural dimensions in mm.

	(a)	(b)	(c)
A_a	130.00	130.00	130.00
A_c	68.00	36.00	130.00
B_a	63.84	65.40	64.96
B_c	44.00	36.00	50.70
$2r_1$		12.00	12.00
a_1	24.00		
b_1	12.00		
$D = \lambda_w/2$	61.20	61.20	61.20
L_a	44.00	44.00	44.00
L_c	4.60	4.60	4.60
T	6.30	6.30	6.30
l_{dt}	4.00	4.00	4.00
t_{dt}	6.00	6.00	6.00
x_{s1}	15.00	15.00	16.00
l_{s1}	36.00	35.00	32.00
d_{s1}	6.00	6.00	8.00
y_{s2}	21.00	15.00	16.00
l_{s2}	48.00	49.00	40.00
d_{s2}	6.00	6.00	8.00

design (b) had circular beam holes and rectangular coupling slots, finally, design (c) had circular beam holes and coupling slots.

In the accelerating mode, the strong magnetic field of the tuned structure provides inter-cell coupling. The coupling cell dimensions are small. This size strongly decreases its frequency as compared to the frequency of the accelerating-cell; therefore to tune it to the accelerating-cell frequency, the dimensions of the coupling cell must be decreased so that it almost intersects the coupling-slot outer limit. For small coupling cells and coupling slots close to the axis, the integral of the surface current induced by this coupling-cell field is small. Since at the slot the magnetic field of the coupling-cell is small compared to the stored energy, it is possible [see formula (3.14)] to obtain a high coupling with only small additional losses.

Table 4.10. Electrodynamic characteristics.

	(a)	(b)	(c)
k_c	15.1%	15.2%	13.2%
$r_{sh \cdot eff}$ [MΩ/m]	40.4	48.9	48.8
G_x	275 m^{-2}	–680 m^{-2}	–650 m^{-2}
G_y	–275 m^{-2}	680 m^{-2}	650 m^{-2}
Q	12400	13400	13000

The electrodynamic characteristics for these designs are summarized in Table 4.10. The focusing gradients with circular beam apertures are of opposite sign to those with rectangular 10×20 mm beam slots.

Calculating the on-axis longitudinal electrical field for the design shown in Fig. 4.12b, we assumed a tuned structure consisting of two coupling cells, one accelerating cell, and two accelerating half-cells terminated by electric walls.

Figure 4.13b shows the dispersion in a tuned structure. We note asymmetry with respect to the $\pi/2$ point which indicates the direct coupling between accelerating cells, a common feature for

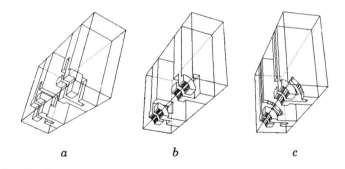

a   b   c

Figure 4.12. Three rectangular biperiodic structures.

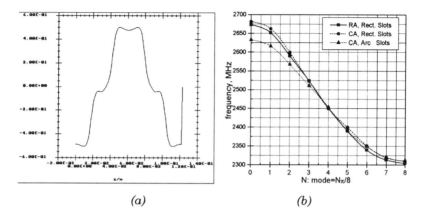

(a)   (b)

Figure 4.13 (a) On-axis E_z and (b) dispersion characteristics.

structures with high coupling; this fact must be taken into account in tuneup.

4.6   BPSs with Coaxial-Radial Coupling Cells

The transverse instability of the beam manifests itself as a beam blow-up. For traveling wave linacs, it has been well reported [119]. For standing wave linacs with biperiodic structures, beam instability has not been studied adequately. This type of instability is an important issue for the considered high-efficiency linac, in continuous-mode race track microtrons [29], and in storage rings [33].

It is well known that high order modes (HOMs) are excited in accelerating structures if their frequency coincides with the bunch repetition rate or is close to this rate. It would be natural to ask about the frequency from which one may no longer take into account the effect of HOMs. For such an estimation, one may resort to the Gaussian representation of a bunch with a temporal structure: $\exp(-t^2/4T_p)$, where 99.5% of the charge is contained in a bunch with length $4T_p$. The induced fields decay as

$\exp(-\omega^2 T_p^2/2)$, therefore, as a criterion for the upper frequency limit we can take [34]

$$f_{max} = \frac{1}{\pi T_p \sqrt{2}}.$$ (4.64)

From the standpoint of transverse instability which shortens (breaks up) a beam pulse, the most dangerous mode is TM_{110} that differs in frequency from the operating mode by approximately 1.7 times. Estimates indicate that suppressing only this mode one can increase the threshold current tenfold [120].

In estimating the effect of HOMs on the process of acceleration, one commonly refers to the loss coefficient, defined in Chapter 1.

Several methods have been reported of how one can increase the threshold current which is inversely proportional to the transverse shunt impedance $R_{sh\perp} = r_{sh\perp}L$, where $r_{sh\perp}$ is the transverse shunt impedance per unit length and L is the length of the accelerating cavity.

One method to increase I_{th} in an on-axis coupled structure consists of shortening the effective length L by detuning the cells in the TM_{110} mode by changing the relative orientation of elements but retaining the frequency of the operating mode invariable [120].

Another way to increase the threshold current is to selectively degrade the quality factor at the frequency of the TM_{110} mode by resonance pumping antennas or cutoff waveguides. These measures reduce $r_{sh\perp}$.

Since the quality factor of coupling cells normally does not exceed $0.2Q$ of accelerating cells, one can vary the coupling coefficient to considerably damp the electromagnetic field induced in the coupling cells in the TM_{110} mode if the beam velocity equals the speed of light.

Good results can be obtained if one uses coaxial or coaxial-radial designs of coupling cells in which the beam does not excite these cells directly. In such BPSs, coupling and accelerating cells are subjected to essentially different conditions and have respectively different frequency spectra. This fact is proved by the data presented in Section 1.2 and experiments on excitation of BPSs with on-axis coupling cells and coaxial cells by a pulse electron beam [121].

The design of BPSs with internal coaxial coupling cells is presented in Fig. 4.14a. The coupling cells are excited in the TM_{010} mode of the coaxial type. The dimensions of these cells are determined without allowance for the effect of coupling slots from the transcendental equation

$$\frac{N_0(kr_1)}{J_0(kr_1)}\frac{J_0(kr_2)}{N_0(kr_2)}=1\,. \tag{4.65}$$

A drawback of these system is a considerably increased external diameter of the body $2B$. For example, coaxial coupling cells were used in the BPS of the microtron in the Institute of Nuclear Physics at Moscow State University [28]; for $r_1 = 12$ mm, $\lambda_0 = 122.4$ mm, $r_2 = 71.2$ mm, the external diameter of the body $2B$ must not be smaller than $2r_2 + 2(0.04\lambda_0) = 152$ mm.

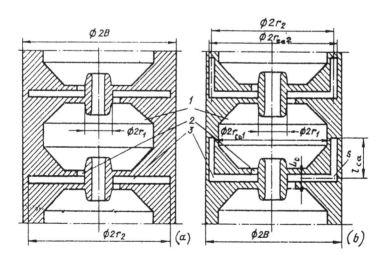

Figure 4.14. BPSs with (a) coaxial an (b) coaxial-radial coupling cells: 1, accelerating cell; 2, coupling slot; and 3, coaxial-radial coupling cell.

In order to improve the size and weight characteristics of BPSs with coaxial coupling cells, designers prefer a configuration in which the external part of the coupling cell is made as a annular recess coaxial with the beam channel. The dimensions of such a coupling cell indicated in Fig. 4.14b are determined from the equality of the input impedances of the coaxial and short-circuited radial waveguides [99]:

$$\frac{Z_0}{2\pi}\ln\frac{r_{ca2}}{r_{ca1}}\tan(kl_{ca}) = -\frac{L_c Z_{02}}{2\pi r_2}\frac{\sin(\theta_2-\theta_1)}{\cos(\psi_2-\theta_1)},\qquad(4.66)$$

where the wave impedance of the radial waveguide is

$$Z_{02}=Z_0\frac{\sqrt{J_0^2(kr_2)+N_0^2(kr_2)}}{\sqrt{J_0^2(kr_2)-N_0^2(kr_2)}}=\frac{Z_0 G_0(kr_2)}{G_1(kr_2)},$$

$$\theta_n=\arctan\frac{N_0(kr_n)}{J_0(kr_n)},$$

$$\psi_n=\arctan\frac{J_1(kr_n)}{N_0(kr_n)}.$$

Recognizing that L_c is selected in the range $0.03\le L_c\le0.04$, and $r_{ca2}/r_{ca1}\cong1$, relation (4.66) can be simplified as

$$Z_0\tan(kl_{ca})\approx-Z_{02}\frac{\sin(\theta_2-\theta_1)}{\cos(\psi_2-\theta_1)}.\qquad(4.67)$$

Here, kr_1 is selected equal to 0.62 from the condition of optimization of the accelerating cell with respect to $r_{sh,eff}$. For this value of kr_1, both sides of equation (4.67) have been worked out in Fig. 4.15. The thickness of the wall between the annual recess and the accelerating cavity is selected from the relation $r_{ca1}-R\cong0.03\lambda_0$. In the considered example, $L_c=4$ mm, $r_{ca1}-r_{ca2}=L_c$ and $r_{ca1}=47$ mm, which yields $kr_2=2.52$. From the plot in Fig. 4.15 we find $Z_{02}(2.52)=1.53Z_0$ and $l_{ca}=1/k=\lambda_0/2\pi=19.4$ mm. The

Figure 4.15. Relative waveguide resistance of a radial waveguide at $kr_1 = 0.62$ (curve 1) and the respective length of the coaxial part of the coupling cell kl_{ca} (curve 2) versus kr_2.

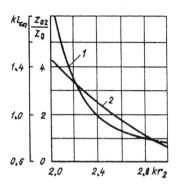

thickness of the partition wall between the annual recess and the outer surface of the body is conditioned by the rigidity consideration $B - r_{ca2} \geq 0.03\lambda_0$. Thus, the outer diameter $2B = 2(44 + 3 + 4 + 4) = 110$ mm for the BPS geometry depicted in Fig. 4.14b, in place of $2B = 152$ mm for the BPS configuration shown in Fig. 4.14a.

In BPSs with coaxial-radial coupling cells, the coupling slot geometry is calculated as follows. The energy stored in the radial part is calculated by the relation for the longitudinal component of the electric field

$$E_{\hat{z}}(kr) = E_0 \frac{G_0(kr)}{G_0(kr_{max})} \frac{\sin[\theta_1 - \theta(kr)]}{\sin[\theta_1 - \psi(kr_{max})]}, \qquad (4.68)$$

where E_0 is the first maximum of E_z at kr_{max} calculated from the condition $\psi(kr_{max}) = \pi/2 + \theta_1$. From equation (4.68) plotted for $kr_1 = 0.62$ in Fig. 4.16 it follows that E_z peaks at $kr_{max} = 1.85$. As can be see from equation (4.68), the integral

$$W_1 = \frac{\varepsilon_0}{2} \int_{v_0} E_z^2(kr)dv$$

cannot be taken in the general form and will have to be computed by numerical integration of the plot in Fig. 4.16. For $\lambda_0 = 0.1224$ m and $L_c = 4$ mm, this integral (in J) is

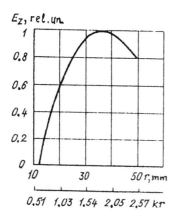

Figure 4.16. Distribution of the electric field in the radial waveguide at $kr_1 = 0.62$.

$$W_1 = \frac{\varepsilon_0}{2} \int_0^{2\pi} \int_0^{L_c} \int_{r_1}^{r_2} E_z^2(kr)\, rd\,rd\varphi\,dz$$

$$= \pi\varepsilon_0 L_c \int_{r_1}^{r_2} E_z^2(kr)\,rd\,r = 10.6 E_0^2 \varepsilon_0 \cdot 10^{-6} \text{ J.}$$

The energy stored in the coaxial part of the coupling cell, W_c, is determined by matching the fields at point s indicated in Fig. 4.14b: $E_{ca}(l_{ca}) = E_z(kr_2)$. Then

$$W_{ca} \approx \frac{\varepsilon_0}{2}\, \pi(r_{ca2} - r_{ca1}) r_2 l_{ca} \left[1 - \frac{\sin(4\pi l_{ca}/\lambda_0)}{4\pi l_{ca}/\lambda_0} \right] E_z^2(kr_2). \quad (4.69)$$

Substituting in this expression the numerical values of parameters determined and $E_z(kr_2) = 0.814 E_0$ yields $W_{ca} = 2.4 \times 10^{-6} \varepsilon_0 E^2$ J. The magnetic field in the radial line short-circuited at $r = r_1$ is defined as

$$H_\varphi(kr) = j\frac{E_0 G_1(kr)}{Z_{02}(kr_{max}) G_1(kr_{max})} \frac{\sin[\theta(kr_1) - \psi(kr)]}{\sin[\theta(kr_1) - \theta(kr_{max})]}. \quad (4.70)$$

Like in an accelerating cell (see Fig. 3.1), in a coupling cell, the magnetic field is a maximum at $kr = 2\pi r_1/\lambda_0$; this coordinate locates the coupling slot cutting area. Using relation (4.70) together with the result of numerical integration of (4.68) and formula (4.69) we find $H_\varphi^2(kr_1)/(W_1 + W_k) = 0.289 \times 10^6/\mu_0$ A^{2}J^{-1}m^{-2}. At $r = r_1$, the magnetic field of an accelerating cell can be computed by programs described in Section 2.2. The respective data for a selected accelerating cell are presented in Fig. 3.1; $H^2(r = 12 \text{ mm})/(W_1 + W_{ca}) = 1.28 \times 10^4/\mu_0$ A^{2}J^{-1}m^{-2}. Now, we use the condition (3.12) of normalization of the magnetic field on coupling slots:

$$H_c(r_{sl})H_a(r_{sl}) = \frac{\mu_0}{2}\sqrt{\frac{H_c H_a^2}{W_c W_a}}$$

$$= 0.5\sqrt{28.9 \cdot 10^4 \cdot 1.28 \cdot 10^4} = 3 \cdot 10^4 \text{ m}^{-3}.$$

This value must be inserted in formula (3.14) for slot dimensions. If we take $t = 4$ mm and $\Delta = 6$ mm, then $Z_{sl}/Z_0 = 0.57$ and, for the given $k_c = -0.03$, we obtain

$$l_{sl} \approx \left\{ -6k_c\left[H_c(r_{sl})H_a(r_{sl})\frac{Z_{sl}}{Z_0}\right]^{-1} \right\}^{1/3} = 21.9 \text{ mm}.$$

The angular span of this slot is about 90° [120].

Thus, the accelerating structure with coaxial-radial coupling cells retains the positive properties of its counterpart with coaxial coupling cells, but has substantially improved weight and size characteristics.

4.7 BPS with Cross Bars. Annular BPS

Accelerating structures with cross bars (SCB) have been studied in detail as a feasible design of an acceleration section for a storage ring [120–123]. Despite its modest transverse dimensions (around 0.4λ), this type of structure has a high shunt impedance and $\beta_{gr} \sim 0.4$.

The positive properties of the SCB can be enhanced by configuring it as a biperiodic structure with cross bars (BSCB) [124] illustrated in Fig. 4.17. Each second spacing between adjacent stems is 5–6 times shorter than others, and the drift tubes are attached to stems at their rim facing the closest stem. This geometry is equivalent to the ratio of the lengths of accelerating and coupling cells $L_a/L_c \approx 3.5$ to 4.0.

This structure has a negative dispersion whose slope depends on the angle ψ between the nearest stems. At some angle ψ, the resonant frequencies of accelerating and coupling cells coincide and the dispersion characteristics has no discontinuities around the $\pi/2$ mode. Evolution of the BSTS dispersion curve with the angle ψ is illustrated in Fig. 4.18.

This BSCB structure is assembled from modules corresponding to sections A–A in Fig. 4.17. It is tuned to the operating frequency of the $\pi/2$ mode with a continuous dispersion characteristic in two steps. The first step includes a preliminary simultaneous tuneup of all cells of the cavity mockup. For this purpose, the angle between the nearest stems in the assembled mockup is varied until the discontinuity in the dispersion curve for the $\pi/2$ mode disappears. If the frequency $f_{1/2}$ is lower than the operating frequency, then the accelerating gap d_g is enlarged by

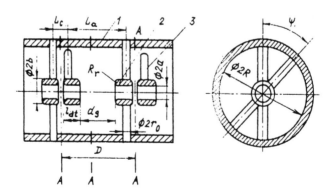

Figure 4.17. Biperiodic stricture with cross bars: *1*, body; *2*, drift tube; and *3*, stems.

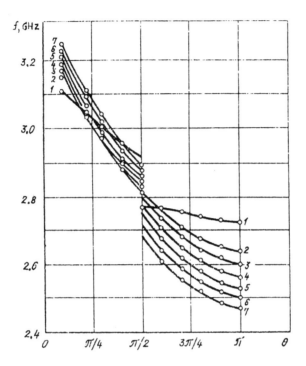

Figure 4.18 Evolution of the dispersion characteristics of a BSCB as for different angles between adjacent stems of coupling cells. Curves 1 to 7 correspond to ψ = 12, 24, 28, 32, 36, 40, and 44°.

cutting off material from the drift tube ends and ψ is varied again until the discontinuity in the dispersion curve disappears. This operation is repeated until the frequency of the $\pi/2$ mode is 2–3 MHz below the desired frequency.

The second step consists of final individual tuneup of mockup cells. A mockup with two reference cells is tuned exactly to $f_{1/2}$, as in the first step, at ψ obtained in this step.

The drift tubes of the reference mockup must be identical accurate to ±0.02 mm. Then, all the remaining cells are tuned in turn to the operating frequency in a mockup of two cells one of which is a reference cell. The relative tuneup accuracy is $\Delta f_0/f_0 \approx 3 \times 10^{-3}$. This tuneup is a straightforward procedure since the derivatives of the $\pi/2$ mode frequency with respect to d_g and ψ are rather large. For example, for $f_{1/2} = 2.8$ GHz, $2a = 8$ mm, $D = 53.2$ mm, $2R = 38.5$ mm, $2r_0 = 7$ mm, $l_{dt} = 17.2$ mm, $2b = 15$ mm, $L_c = 11.5$ mm, and $\psi = 24°$, we have $\partial f_{1/2}/\partial d_g = 60$ MHz/mm and $\partial f_{1/2}/\partial \psi = 3.4$ MHz/deg. The measured characteristics of such a BSCB are as follows: $f_{1/2} = 2.8$ GHz, $r_{\text{sh,eff}}/Q = 4.8\pm0.1$ kΩ/m, $Q = (6.8\pm0.3) \times 10^3$, $k_c = -0.25$, and $k_{ov} = 6.6$ for the edge rounding radius of drift tubes $R_r = 1.7$ mm. At an rf power of 1 MW, the accelerating section built around a BSCB provides for $\Delta W = 3.5$ MeV of beam energy at a zero beam load. At the frequency $f_{1/2}$, the internal diameter of a BSCB is equal to $2R = 0.36\lambda_0$.

Investigations indicate that the third passband of a BSCB, formed by the TM_{01} mode, has the impedance and weight and size characteristics similar to those of BPSs with internal coupling cells. However, the coupling coefficient of a BSCB excited in the TM_{01} mode ($k_c \approx 0.4$) corresponds to a DAW structure whose outer diameter is by 1.4–1.5 times larger.

A shortcoming of slotted biperiodic structures (with magnetic coupling) arises in that an increase of the coupling coefficient would considerably decrease the effective shunt impedance per unit length:

$$r_{\text{sh,eff}}(k_c) = r_{\text{sh,eff}}(k_c = 0)\left(1 - 9.0|k_c|^{3/2}\right). \qquad (4.71)$$

A biperiodic structure with a weaker dependence of the effective shunt impedance on the coupling coefficient is presented in Fig. 4.19a; its equivalent circuit is shown in Fig. 4.19b. This structure is referred to as an annual BPS [125]. The principal difference of this structure from the known configurations lies in that the entire structure is an annual, rather than linear, chain of coupled cells. Its equivalent circuit is characterized by lumped parameters R_p, L_p, and C_p. Assuming identical coupling coefficients between the circuits $k_p = k_c$, electric oscillations in circuit p may be described by

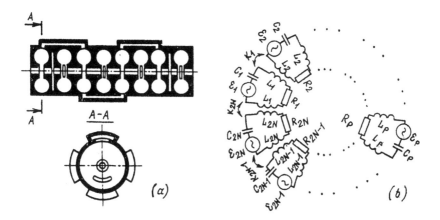

Figure 4.19 (a) Annular BPS, and (b) its equivalent circuit.

$$I_p = i_p\left[1 + \frac{f_0}{jfQ} + \frac{f_0^2}{f^2}\right] + \frac{k_c}{2}(i_{p-1} + i_{p+1}), \qquad (4.72)$$

where $I_p = E_p/(4\pi jfL_p)$ is the excitation current of cell p, E_p is the emf introduced in this circuit, f is the frequency of an external emf, i_p is the current in the circuit, f_p is the resonant frequency of circuit p, and $Q = 4\pi f_0 L_p/R_p$ is the quality factor of circuit p.

Solving the system of $2N$ equations (4.72), set up for a structure with an even number of accelerating cells, for currents i_p yields the eigenfunctions of the circuit $i_{pn} = A\cos(\theta_p)\exp(2j\pi f_p t)$ and a dispersion equation (4.46), however, this time, $\theta = \pi n/N$ for $n = 0, 1, 2, ..., 2(N - 1)$ rather than $\theta = \pi n/2(N - 1)$ as with the linear BPS. Consequently, the number of modes in a linear BPS and in an annular BPS is respectively equal to $2N - 1$ and $N + 1$. The frequency separation of the $\pi/2$ mode from adjacent modes is

$$\Delta f_{1/2}^{(an)} \approx \frac{\pi k_c}{4(N-1)}f_0 \text{ and } \Delta f_{1/2} \approx \frac{\pi k_c}{2N}f_0 \qquad (4.73)$$

where the superscript "an" refers to the annular structure.

This expression suggests that, when the coupling coefficient of the annular structure is identical with that of the linear system and N grows without bound, the annular structure is excited in a half as large number of modes and, consequently, has twice as large frequency separation of the operating and adjacent modes. With an identical configuration of accelerating cells and an equal coupling coefficient, the annular structure will have a higher effective shunt impedance than the usual linear structures.

The annular biperiodic structure has Ω-shaped accelerating cavities, and coupling cells are made as prismatic cells resonant in the TE_{101} mode. In order to reduce the transverse size of the structure, coupling cells are bent parallel to the external cylindrical surface of accelerating cells. Two extreme accelerating cells are coupled via usual cylindrical cells resonant in oscillations of the TM_{010} mode.

The annular BPS may have any number of accelerating cells. The spatial arrangement of coupled cells in this structure is such that the relation $\theta = \pi n/N$ holds only for an even N. For an odd N, the total phase advance in an annular chain is $\Sigma_i \varphi_i = 2\theta N - \pi$. This phase advance musts be $\Sigma_i \varphi_i = 2\pi n$. For an odd N, this condition yields $\theta N = 2(n + 1)\pi/2N$, where $n = 0, 1, 2, ..., N - 1$.

Calculations of the coupling slot geometry for the annular BPS have a number of specific features associated with the configuration of coupling cells. The nomenclature of slot dimensions and a coordinate system tied with the coupling cell are illustrated in Fig. 4.20. The transverse dimension of the cell is measured along the arc A over the body of accelerating cells. Using the known relations for a prismatic resonator excited in the TE_{101} mode, we obtain

$$
\left.
\begin{aligned}
H_x &= j\frac{E_0}{Z_0}\frac{A}{\sqrt{A^2 + L_c^2}}\cos\frac{\pi x}{A}\sin\frac{\pi z}{L_c}, \\
\\
W &= \varepsilon_0 ABL_c\, E_0^2\big/8, \\
\lambda_0 &= 2A\,L_c\big/\sqrt{A^2 + L_c^2}.
\end{aligned}
\right\}
\tag{4.74}
$$

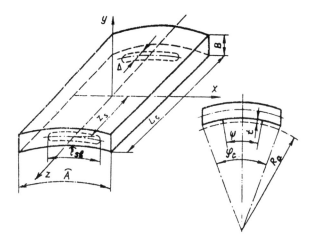

Figure 4.20. Coordinate system tied with a coupling cell of an annular biperiodic structure.

Then

$$H_c^2 = \frac{4\overline{\cos}^2\left(\dfrac{\pi}{A}l_{sl}\right)\sin^2\left(\dfrac{\pi}{L_c}l_{sl}\right)}{\left(1+\dfrac{L_c^2}{A^2}\right)ABL_c}, \qquad (4.75)$$

where $A = (R_a + t)\varphi$, and the average cosine is

$$\overline{\cos}\left(\frac{\pi}{A}l_{sl}\right) = \frac{1}{l_{sl}}\int_{-l_{sl}/2}^{l_{sl}/2}\cos\left(\frac{\pi}{A}z\right)dz = \frac{\sin\left(\dfrac{\pi}{2A}l_{sl}\right)}{\dfrac{\pi}{2A}l_{sl}}.$$

As an example, we perform a calculation of coupling cells for the RELUS-4 system with $\lambda_0 = 107.2$ mm, $L_a = 2D + \Delta = 113.2$ mm, $B = 4$ mm, $z_{sl} = 53.6$ mm, $t = 4$ mm, $\Delta = 6$ mm, and, as follows from (3.11), $Z_{sl}/Z_0 = 0.57$.

Disregarding the perturbation due to slots, we find the width A of a coupling cell from the last relation of the system (4.74) as $A = 60.9$ mm. Since the length of slots is unknown so far, we determine H_c^2 from (4.74) in the first approximation by letting $\overline{\cos}(\pi z_{sl}/A) = 1$. Then, $H_c^2 = 3 \cdot 25 \cdot 10^4$ m^{-3}. Since the coupling slot has the coordinate $r_{sl} = R_a$ with respect to the accelerating cell, relation (3.13) yields $H_a^2 = 0.38 \cdot 10^4$ m^{-3}. Using the obtained values for H_c^2, H_a^2, and Z_{sl}/Z_0, we find, for $k_c = -0.04$, the length of the coupling slot $l_{sl} = 33.6$ mm. Now, we refine the value of H_c^2 with $\overline{\cos}^2(\pi z_{sl}/A) = 0.773$ and calculate a new value of $l_{sl} = 36.6$ mm.

The slot is made with rounded ($R_{r,sl} = A/2$) edges, so that its full length will be $l_{sl,r} = 38.0$ mm. In tuning the accelerating section of RELUS-4, we obtained $l_{sl,r} = 37.4$ mm. To ensure a high acceleration rate, we assumed $d_g/\lambda_0 = 0.74$. This value produces [126] an insignificant reduction of the shunt impedance as compared to the optimal value at $d_g/\lambda_0 \approx 0.6$. The overvoltage coefficient reduces significantly, thus increasing the electrical strength of the structure. From experiments we obtained $r_{sh,eff}/Q = 5.0$ kΩ/m, $Q = 1.4 \times 10^4$, and $k_{ov} = 2.8$. At $k_c = 0.04$, $r_{sh,eff} = 65$ MΩ/m, and the structure has the maximum possible acceleration rate which is by 1.7 times higher than that in a BPS with the optimal ratio $d_g/\lambda_0 = 0.6$. The maximum transverse size of the accelerating system, determined by the top outer walls of the coupling resonators, is $0.95\lambda_0$.

Table 4.11 Characteristics of different BPSs.

Structure	$r_{sh,eff}$ MΩ/m	Q_0	k_c	k_{ov}	$2R_1$, cm
BPS(an)	66	14500	0.05	3.4	16.0
BPS(B)$_{opt}$ (Fig. 4.14)	72	13500	0.03	3.8	11.5
BSCB(TEM)	33	6800	0.25	6.6	6.0
BSCB(TM$_{01}$)	65	15000	0.4	4.1	12.5
Ring-type	66	14500	0.08	2.8	12.0

Thus, at $|k_c| \geq 5\%$, the annular BPS structure obviously outperforms its counterparts in the shunt impedance while preserving a compact transverse size.

Table 4.11 summarizes the electrodynamic characteristics of an on-axis coupled structure with annular coupling cells (an), BSCBs resonant in the TEM mode and TM_{01} mode, and an annular biperiodic structure. Measurements were conducted at 2.8 GHz. The right column shows the outer diameter of the structure measured over the water cooling hood.

4.8 Optimization of Electrodynamic Range

For given rf power parameters, the desired levels of beam energy and current can be achieved with different combinations of the main linac electrodynamic characteristics. Reference data compiled in our handbook [3] present wide possibilities for analysis of different designs of accelerating sections based on DLWs with constant or variable dimensions. A feasible initial design is normally chosen subject to some criteria, such as electronic performance, weight and size characteristics, beam parameter stability, and technological simplicity.

One of the most popular DLW optimization techniques consists of choosing a design with the highest value of $E_0\lambda_n P^{-1/2}$ at the minimal values of attenuation and overvoltage coefficients. It is not hard to demonstrate that the possibility of simultaneous optimization of the waveguide with respect to $E_0\lambda_n P^{-1/2}$ and electrical strength at a given group velocity ($\beta_{gr,n}$ = constant) is rather limited.

Let us consider the plot in Fig. 4.21 showing $E_0\lambda_{1/2}P^{-1/2} = F(\beta_{gr,1/2})$ for $\beta_{ph,1/2} = 1$ and $0.02 \leq t/\lambda_{1/2} \leq 0.095$. It indicates that, for $\beta_{ph,1/2}$ = constant, the parameter $E_0\lambda_{1/2}P^{-1/2}$ maximizes for an iris whose relative thickness $t/\lambda_{1/2}$ tends to zero. Clearly, this configuration will also have the minimal attenuation. On the other hand, for a rounded iris with a rounding radius $R_r = t/2$, the overvoltage coefficient, k_{ov}, decreases as $t/\lambda_{1/2}$ increases. For

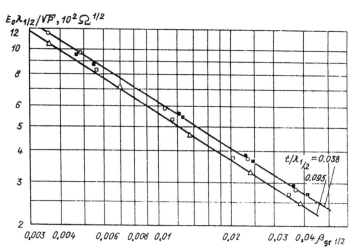

Figure 4.21 Variation of $E_0\lambda_{1/2}P^{-1/2}$ as a function $\beta_{gr,1/2}$ for two values of $t/\lambda_{1/2}$.

example, at $\beta_{gr,1/2} = 0.014$, an increase of the waveguide electric strength due to passing from $t/\lambda_{1/2} = 0.038$ to $t/\lambda_{1/2} = 0.095$ leads to a decrease of the parameter $E_0\lambda_{1/2}P^{-1/2}$ from 500 to 450 $\Omega^{1/2}$ and an increase of the attenuation parameter from 6.6×10^{-3} to 7.4×10^{-3} $m^{1/2}$. For these options, the experimental values of k_{ov} are 1.84 and 1.69, respectively. As a result, an increase of the electrical strength by 10% reduces the efficiency of rf power conversion into the electron energy (R_{sh} decreases from 57 to 42 MΩ/m at $\lambda = 0.107$ m) and the positive effect of this optimization is doubtful.

The scope of DLW optimization with respect to the electrical strength could be expanded if one departs from the traditional plane boundary of the iris over its entire surface except the aperture. An optimization scheme including choice of an initial waveguide geometry with cylindrical cells with $R_r = t/2$ and variation of the iris profile by changing its boundaries into the exterior of the cells can somewhat reduce the group velocity, since it increases the effective iris thickness. In this setting, a decrease of the quality factor caused by an increase of $t/\lambda_{1/2}$ can be offset by bringing the cell profile closer to an Ω shape.

We chose the cell boundary profile with a program package OPTIM [61] which performs numerical computations only with 0 and π modes. The suitability of this treatment for our problem (optimization in the π/2 mode) was proved experimentally. For experiments, we used resonance DLW mockups of three cells with terminating half-cells. The irises of this mockup had $t/\lambda_{1/2} = 0.038$ and $R_r/t = 1/2$.

The distribution of the normal component of the electric field over the iris was measured by an automated broadband test bench with a 1.5×1.5 mm plate dielectric probe. This probe was moved so that its 1.5×1 mm face slid over the iris. Figure 4.22a

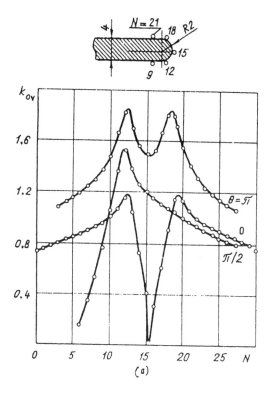

Figure 4.22. Overvoltage coefficient, k_{ov}, on the surface of irises with different profiles in 0, π/2, and π modes.

presents the results of the measurements of $k_{ov} = E_{s,max}/E_{z,max} = F(N)$, where N is the probe coordinate, $E_{s,max}$ is the maximum rf field on the surface of the iris, and $E_{z,max}$ is the maximum strength of the electric field on the mockup axis. The probe travel between points N and $N+1$ is equal to 0.74 mm. In 0, $\pi/2$, and π modes, the electric field is concentrated at one and the same point *12* lying near the point where the toroidal surface of the aperture matches the plane of the iris. The overvoltage coefficient peaks at the π mode.

The initial design of optimization corresponded to a DLW with

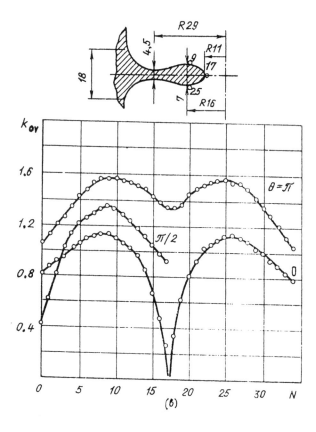

Figure 4.22. Continued

the parameters $\theta = \pi/2$, $a/\lambda_{1/2} = 0.102$, $t/\lambda_{1/2} = 0.038$, $\lambda_{1/2} = 0.1072$ m, $\beta_{ph,1/2} = 1$, and $a/b = 0.26$. A numerical optimization of this initial setting with respect to k_{ov} yielded a value of k_{ov} which is lower than the initial value by 15% In accordance with the calculated data we made a resonance mockup with the iris profile depicted in Fig. 2.22b. This figure also shows $E_{s,max}/E_{z,max}$ measured over the contour of the iris. A comparison of diagrams at (a) and (b) indicates that, in the 0 mode, k_{ov} remain invariable, whereas in the π mode, $k_{ov}^{(2)}/k_{ov}^{(1)} = 0.86$, which almost coincides with the calculated data. Measurements of the initial and optimized profile were conducted with one and the same probe and the error in determining $k_{ov}^{(2)}/k_{ov}^{(1)}$ was within ± 0.01. The measured electrodynamic characteristics of the initial and optimized mockups are summarized in Table 4.12.

This table lists the parameter $E_0\lambda_{1/2}P^{-1/2}$ for the accelerating harmonic of the electric field and $k_{ov,0}$ corresponds to $E_{s,max}$ normalized by this value. Data presented in this table can be used to estimate the advantages of the optimized waveguide over the prototype. If we assume $E_{s,max} = 300$ kV/cm, then, at the outset of the initial accelerating section, $E_0^{(1)} = 16.3$ MV/m, and $E_0^{(2)} = 18.3$ MV/m. To achieve such fields, these designs require rf power levels of $P^{(1)} = 15$ MW and $P^{(2)} = 15.3$ MW, respectively. Then, for a beam current of $I = 4$ mA, $W^{(1)} = 44.1$ MeV and $W^{(2)} = 47.1$ MeV.

Table 4.12 Characteristics of the initial and optimized resonance mockups

Mockup	$\beta_{gr,1/2}$	$E_{z,max}\lambda_{1/2}P^{-1/2}$ $\Omega^{1/2}$	$E_0\lambda_{1/2}P^{-1/2}$ $\Omega^{1/2}$	r_{sh}/Q kΩ/m
initial (1)	0.0148	545	451	4.47
optimized (2)	0.0122	609	502	4.64

Mockup	Q	α, m^{-1}	k_{ov}	$k_{ov,0}$
initial (1)	11000	0.18	1.52	1.84
optimized (2)	11400	0.21	1.35	1.64

Thus, the optimized profile of the iris increases the energy by 7% when the rf power increases by 2%. Under the circumstances, the electron efficiency increases insignificantly: $\eta^{(1)} = 11.8\%$ and $\eta^{(2)} = 12.3\%$.

5

H-TYPE
RESONATORS

5.1 Background

A uniform H-type cavity with an axially symmetric beam channel and cylindrical body is schematically represented in Fig. 5.1. The nomenclature used in this figure as follows: R denotes the radius of the body, L the length of the cavity, l_t the length of a drift tube, d_g the length of an accelerating gap, $D = l_t + d_g$ the period of the structure, b the radius of drift tubes, and r_0 the radius of stems. The figure also shows a tied cylindrical coordinate system r, φ, z and an approximate distribution of the magnetic field over the cavity cross section at $z = 0$.

An analytical solution of the wave equation with boundary condition for such an H-type resonator may be written in the form of series containing an expansion in terms of the eigenfunctions in partial domains formed by the body in the gap segment d_g, body,

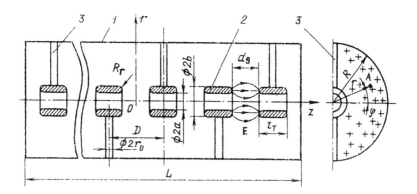

Figure 5.1. H-type resonator with axially symmetric beam channel: 1, body; 2, drift tubes; and 3, stems.

and drift tubes, and also the body, drift tubes, and a stem. Recognizing the axial asymmetry of the last domain one may conclude that the calculation of the coefficients in such series and their summation will be an almost unsolvable problem. These difficulties can be avoided by the method of equivalent circuits or the method of cavity analogue.

A cavity analog of an H-type resonator is chosen on the basis of known experimental data about the topography of the magnetic field in such systems [27, 128]. Typical profiles of $H_z(r, \varphi, z)$ for various coordinates φ are presented in Fig. 5.2. At some distance from the stems, the lower passband has a magnetic field profile similar to a TE_{111} mode in a cylindrical or coaxial cavity. The operating profile of these oscillations corresponds to one variation of the longitudinal H component along the cavity.

Thus, as a cavity analogue we choose a cylindrical or coaxial resonator with a homogeneous dielectric filling medium and excitation in the TE_{111} mode. The relative dielectric permittivity of a conditional dielectric ε_1 must provide the equality of the resonance frequencies for the cavity-analog and the prototype with like dimensions of the body. For an analogue in the form of a cylindrical cavity, the magnetic field is

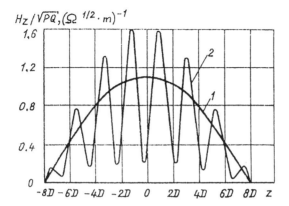

Figure 5.2 Distribution of reduced axial magnetic field over the surface of H-type resonator for (1) $\varphi = 0$ and (2) $\varphi = 84°$. Measurements were performed with a ring probe with a step $\Delta z = 1$ mm, $D = 12.6$ mm.

$$
\left.\begin{aligned}
H_z(r,\varphi,z) &= BJ_1(k_{cr}r)\cos\varphi\,\cos(\pi z/L), \\
H_\varphi(r,\varphi,z) &= -B\frac{k_z}{k_{cr}^2}\frac{1}{r}J_1(k_{cr}r)\sin\varphi\,\sin(\pi z/L), \\
H_r(r,\varphi,z) &= -B\frac{k_z}{k_{cr}}J_1'(k_{cr}r)\cos\varphi\,\sin(\pi z/L).
\end{aligned}\right\}
\tag{5.1}
$$

The wave numbers $k_z = 2\pi/\lambda = \pi/L$, $k_{cr} = \mu_{11}/R$, and $k_1 = 2\pi\varepsilon_1^{1/2}/\lambda_0$ are related by

$$
k_z^2 = k_1^2 - k_{cr}^2 .
\tag{5.2}
$$

For a coaxial cavity analogue resonant in the TE_{11} mode, the magnetic field must be written as

$$
H_z(r,\varphi,z) = B\left[J_1(k_{cr}r) - \frac{J_1'(k_{cr}b)}{N_1'(k_{cr}b)}N_1(k_{cr}r)\right]\cos\varphi\,\cos\left(\frac{\pi z}{L}\right)
\tag{5.3}
$$

where $N(k_{cr}r)$ is the first order Neumann function.

The critical wave number k_{cr} is determined from the transcendental equation

$$J_1'(k_{cr}b)N_1'(k_{cr}R) = N_1'(k_{cr}b)J_1'(k_{cr}R).$$

For any ratio R/b, the solutions to this equation are in the range $1 \le k_{cr}R \le 1.84$. In practice, all known configurations of H-type resonators lie in the range $20 \ge R/b \ge 10$; for $R/b = 20$, $k_{cr}R = 1.83$, and for $R/b = 10$, $k_{cr}R = 1.80$.

Figure 5.3 shows the distribution of the longitudinal magnetic field corresponding to a simple cylindrical and coaxial cavity with $R/b = 15$ worked out with formulas (5.1) and (5.3). An essential difference of the magnetic fields is observed for $r \le 0.2R$, where, however, the field is significantly weaker than elsewhere. It should be obvious that the volume and surface integrals of the magnetic field of cylindrical and coaxial cavity analogues will be almost identical, therefore one should use the simpler expressions (5.1).

Replacement of a discrete capacitive load of the cavity by a homogeneous dielectric is justifiable when $D/\lambda_{wg} \ll 1$. Real H-cavity designs used in ion linacs have $D/\lambda_{wg} = 0.05$, which is a good approximation to the limiting case.

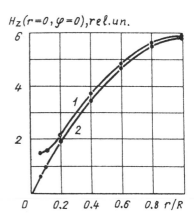

Figure 5.3. Distribution of the longitudinal magnetic field at $\varphi = 0$ in (1) cylindrical and (2) analogues of an H-type resonator.

The concept of H-type resonators is as follows. The stems of adjacent drift tubes together with the connecting arc of the body form a circuit loaded onto capacitance C_0 formed by the drift tubes. This loop is pierced by the alternating magnetic field $H_z(r, \varphi z)\sin\omega_0 t$. By the law of electromagnetic induction, the voltage across this capacitance is

$$U = -\frac{d\Phi}{dt}\bigg|_S = -\mu_0\omega_0 \int_S H_z(r,\varphi,z)ds, \qquad (5.4)$$

where Φ is the flux of magnetic induction through the circuit with area S, $\mu_0 = 1.26 \times 10^{-6}$ H/m, and $c = 3 \times 10^8$ ms^{-1}.

Assuming that the capacitive reactance of the circuit x_C is much larger than its inductive reactance x_L and resistance r_a, the amplitude of a current in this circuit is

$$I = 2U\omega_0 C_0, \qquad (5.5)$$

where it is taken into account that the load in this circuit is formed by two accelerating gaps. For typical values of $\omega_0 = 10^9$ 1/c and $C_0 = 10^{-12}$ F, $x_L = (\omega_0 C_0)^{-1} = 10^3$ Ω and indeed $x_C \gg x_L + r_a$.

The shunt impedance and Q-factor of the H-type resonator can be estimated by calculating the voltage between the drift tubes and current in stems by formulas (5.4) and (5.5).

The size of the body and resonance frequency can be determined if the relation between the parameters of the equivalent dielectric filler and the real capacitive load of the H-cavity is known.

5.2 Estimating of the Size of the Body

The size of the body of an H-type resonator can be estimated by the method of equivalent circuits. The equivalent inductance and capacitance of a round waveguide resonant in the H_{11} mode can be calculated as [129]

$$L_{eq} = 2\mu_0 R^2/(\mu_{11}^2 L), \quad C_{eq} = \varepsilon_0 \varepsilon_1 L/2, \tag{5.6}$$

where $\mu_{11} = 1.841$ is the first root of the equation $J_1' = 0$.

We assume that the waveguide is loaded predominantly by the capacitances of gaps C_0 and drift tubes C_1. Then, the equivalent capacitance of the waveguide with respect to the body is

$$C_{eq}^{(H)} \approx C_{eq} + \frac{C_0 L}{D} + \frac{C_1 L l_T}{D} = \frac{\varepsilon_0 L}{2}\left[1 + \frac{2}{\varepsilon_0 D}(C_0 + C_1)\right], \tag{5.7}$$

and the relative permittivity of a conditional dielectric can be obtained from equations (5.6) and (5.7) as

$$\varepsilon_1 = 1 + \frac{2}{\varepsilon_0 D}(C_0 + C_1). \tag{5.8}$$

Using this expression with formula (5.2) we obtain the resonance wavelength of the H-type cavity

$$\lambda_0 \approx \frac{2\pi R}{\mu_{11}}\sqrt{\frac{\varepsilon_1}{1+(\pi R)^2/(\mu_{11}L)^2}} = \frac{2\pi R}{\mu_{11}}\sqrt{\frac{1+2(C_0+C_1)/(\varepsilon_0 D)}{1+2.91 R^2/L^2}} \tag{5.9}$$

The capacitances between adjacent drift tubes and between drift tubes and the body can be determined by solving the respective electrostatic problems. If we resort to reported data [130] and the known relation for the capacitance of a cylindrical capacitor, then for a H-type resonator with axially symmetric beam channel and cylindrical body we have

$$\left.\begin{aligned}
C_0 &\approx 3 \cdot 10^{-11}(a+b)\left\{1+\frac{b-a}{d_g}+\frac{l_T}{2\pi d_g}\left[\ln\left(1+\frac{2d_g}{l_T}\right)+1\right]\right\}, \\
C_1 &\approx 2\pi l_T \frac{\varepsilon_0}{\ln(R/b)}.
\end{aligned}\right\} \tag{5.10}$$

In these expressions, all variables are in the SI system of units. A proof of the accuracy of formulas (5.9) and (5.10) can be obtained from Table 5.1 which summarizes experimental and calculated data borrowed from Avrelin et al. [128] and Hamming [131] (last column). The capacitances C_0 and C_1 have been calculated by averaging the of outer radius of drift tubes b over the number of accelerating gaps N.

The calculated data indicated in this table exceed the resonance frequency value by 2–3%. This fact may be attributed to the fact that expression (5.8) does not take into account the capacitance of the stem with respect to the body. On the other hand, it would not be reasonable to include this capacitance in the approximate model. Equation (5.9) is suitable also for estimating the eigenfrequency of an H-type resonator with a nonuniform beam channel after averaging the dimensions of its drift tubes.

In the case of H-type resonators, experimental reference data for more accurate dimension analysis are also scarce as in the

Table 5.1 Verification of the H-type resonator theory.

	Designs				
L, m	0.603	0.603	0.603	0.078	1.02
R, m	0.19	0.19	0.19	0.034	0.25
D, mm	25	25	25	6.0	0.25
t_{T}, m	15	15	15	4	15
a, m	5	5	5	1.6	5
b, m	10	12.5	15	3.1	20–30
d_{g}, mm	10	10	10	2	12.5
C_0, pF	0.873	1.15	1.47	0.33	0.95
C_1, pF	0.284	0.31	0.33	0.092	0.28
ε_1	11.44	14.2	17.3	16.12	11.1
f_0^{calc}, MHz	155	139	126	784	114
f_0^{exp}, MHz	153	137	123	767	109

symmetric H-type resonator shown in Fig. 5.1, the relative radius of the body is defined by the functional

$$\frac{R}{\lambda_0} = F\left(\frac{L}{\lambda_0}, \frac{a}{\lambda_0}, \frac{b}{\lambda_0}, \frac{l_T}{\lambda_0}, \frac{r_0}{\lambda_0}, \frac{R_r}{\lambda_0}, \frac{D}{\lambda_0}\right), \tag{5.11}$$

where R_r is the rounding radius of a drift tube. The phase velocity corresponds to the π mode, therefore $\beta_{ph} = 2D/\lambda$. For ion accelerators using variable rf focusing, the parameters l_T/λ_0, D/λ_0 and r_0/λ_0 can vary in a wide range. For example, in the Uragan-2 linac [47], $0.01 \le D/\lambda_0 \le 0.04$ and $0.005 \le l_T/\lambda_0 \le 0.025$. Thus, as in the case of BPSs, the necessary reference data for H-type resonator design can be obtained by calculation with the use of a limited number of experimental tests.

Despite their simple structure, formulas (5.9) and (5.10) pose certain difficulties in calculations. For example, the full expression for the functional (5.11) is

$$\frac{R_0}{\lambda_0} = \frac{\mu_{11}}{2\pi} \sqrt{\frac{1 + \left(\frac{\pi}{\mu_{11}}\right)^2 \left(\frac{R}{L}\right)^2}{1 + 4\pi\left\{0.54\frac{a}{D}\left(1+\frac{b}{a}\right)\left[1+\frac{a}{D}\frac{\frac{b}{a}-1}{1-\frac{l_T}{D}}+\frac{1}{2\pi}\frac{l_T}{D}\left(\ln\frac{2-\frac{l_T}{D}}{\frac{l_T}{D}}+1\right)\right]+\frac{\frac{l_T}{D}}{\ln\frac{R}{b}}\right\}}}$$

$$= \frac{\mu_{11}}{2\pi}\sqrt{\frac{1 + 2.91 R^2/L^2}{1 + 2(C_0 + C_1)/(\varepsilon_0 D)}} \tag{5.12}$$

This solution to this equation is presented in Fig. 5.4. In evaluations of the body radius, the inputs are the calculated geometry data for the beam channel: a/λ, λ, a/b, and l_T/D. For practical calculations one must have a set of plots for different R/b in the range $12 \le R/b \le 20$ and the final solution will correspond to a plot whose R/b parameter will comply with the given radius of

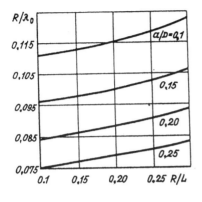

Figure 5.4. Plots to determine the operating wavelength of an axially symmetric H-cavity at b/a = 2.5, l_T/D = 0.5, and R/b = 15.2.

C_1, pF	ε_1	f_0^{calc}, MHz	f_0^{exp}, MHz
0.284	11.44	155	153
0.31	14.2	139	137
0.33	17.3	126	123
0.092	16.12	784	767
0.28	11.1	114	109

drift tubes b. For a nonuniform H-type resonator, the body radius is calculated with averaged dimensions of the beam channel.

Expression (5.12) can be tabulated using the curves plotted in Fig. 5.5 which allow one to determine the capacitance of the axially symmetric gap. Figure 5.6 shows plots to determine the lateral capacitance of a gap.

With reference to Figs. 5.5 and 5.6, the radius of the body is calculated in three steps. First equation (5.12) is solved for the found C_0 on the assumption $C_1 = 0$ and $R/L \to 0$. With the found value of R_{cr} we can determine the first approximation for the lateral capacitance from Fig. 5.6. With the found capacitance

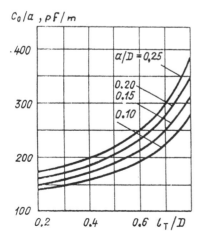

Figure 5.5. Curves to determine the capacitance of the cylindrical accelerating gap at b/a = 2.5.

Figure 5.6. Determining the lateral capacity (C_1/l_T) of the cylindrical gap and the capacity of the grid gap $(C_0/l_{n\Sigma})$.

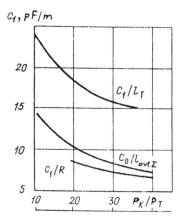

$C_0 + C_1$ we determine a refined value of R. The refined value of R_1, in turn, is used to refine the lateral gap capacitance C_1 and perform the final evaluation of the body radius. In practical calculations of the radius of the body, one may resort to the relations for the derivatives of the resonance frequency with respect to the body radius and with respect to the outer radius of drift tubes

$$\frac{\partial f_0}{\partial R} = -\frac{f_0}{R}\left(1 - \frac{2.91R^2/L^2}{1+2.91R^2/L^2}\right),\qquad(5.13)$$

$$\frac{\partial f_0}{\partial b} = -f_0\frac{\partial C_0/\partial b + \partial C_1/\partial b}{\varepsilon_0 D + 2(C_0 + C_1)}.\qquad(5.14)$$

At $R = $ constant, the derivative for the lateral capacitance is obtained by differentiating (5.10):

$$\frac{\partial C_1}{\partial b} = \frac{C_1}{b\ln(R/b)}.$$

In order to determine the derivative $\partial C_0/\partial b$, Fig. 5.7 presents curves plotted for $b/a = 2.5$. The derivatives (5.13) and (5.14) enable one to do without iterations in solving equation (5.12).

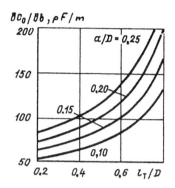

$\partial C_0/\partial b$, pF/m

Figure 5.7. Derivative of the capacitance of a cylindrical gap with respect to the outer radius of drift tube as function of their relative dimensions for $b/a = 2.5$.

5.3 Impedance Characteristics

In Section 5.1 we demonstrated that the use of a coaxial resonator resonant in the TE_{111} mode as a cavity analogue almost does not affect the volume and surface integrals of the magnetic field and therefore one may safely use the simpler representation (5.1). If the magnetic field is specified, the expression for stored energy

$$\frac{PQ}{\omega_0} = \frac{\mu_0}{2} \int_V H^2(r,\varphi,z)\,dv \qquad (5.15)$$

can be written as a magnetic field parameter

$$\xi_M = \frac{H_z(r,\varphi,z)}{\sqrt{PQ}} = \frac{2\mu_{11}J_1(k_{cr}r)\cos\varphi\,\cos(\pi z/L)}{\pi R J_1(\mu_{11})}$$

$$\times \left\{ \frac{\lambda_0}{Z_0 L(\mu_{11}^2 - 1)\left[1 + \pi^2/(k_{cr}L)^2\right]} \right\}^{1/2}. \qquad (5.16)$$

Using this parameter and the low (5.4) we can determine the voltage across the accelerating gap centered at z_n:

$$U = -\frac{2\pi c \mu_0}{\lambda_0} \int\limits_0^R \int\limits_{-\pi/2}^{\pi/2} H_z(r,\varphi,z)r\,dr\,d\varphi = \frac{8R\cos(\pi z_n/L)}{\mu_{11}J_1(\mu_{11})}$$

$$\times \sqrt{\frac{Z_0 PQ}{\lambda_0 L(\mu_{11}^2 - 1)\left[1 + (\pi R)^2/(\mu_{11}L)^2\right]}} \left[\int\limits_0^{\mu_{11}} J_0(x)dx - J_1(\mu_{11})\right]. \quad (5.17)$$

Substituting $\mu_{11} = 1.841$, $J_1(\mu_{11}) = 0.5819$, and

$$\int\limits_0^{\mu_{11}} J_0(x)dx = 1.383$$

in (5.17), we pass to the gap voltage parameter

$$\xi_u(z_n) = \frac{U(z_n)}{\sqrt{PQ}} = \frac{3.87\sqrt{Z_0}R}{\sqrt{L\lambda_0\left(1 + 2.91\dfrac{R^2}{L^2}\right)}} \cos\frac{\pi z_n}{L}. \quad (5.18)$$

Summing this expression over the number of accelerating gaps N, we obtain the shunt impedance of the H-type cavity in the form

$$\frac{R_{sh}}{Q} = \frac{1}{PQ}\left[\sum\limits_{-N/2}^{N/2} \xi_u(z_n)\right]^2 = \frac{U^2(z_n = 0)}{P}\left[\sum\limits_{-N/2}^{N/2} \frac{\cos(\pi n)}{N}\right]^2$$

$$\approx 5.64 \cdot 10^3 \frac{R^2 N^2 k_g^2 k_{vd}^2}{\lambda_0 L(1 + 2.91\,R^2/L^2)}, \quad (5.19)$$

where k_g is the gap efficiency coefficient and $k_{vd} = 0.64$ is the coefficient reflecting the voltage distribution over the gaps. For a resonator tuned to a uniform electric field, $k_{vd} = 1$.

Formula (5.19) gives the parameter of the electric field at the center of accelerating gaps $\xi_e(z_n) = E_z(z_n)/\sqrt{PQ} = \xi_u(z_n)\hat{E}$, where $\hat{E}$ is the normalization factor.

The maximum Q-factor of the H-cavity, Q_{max}, corresponds to the quality factor of a cylindrical cavity excited in the TE_{111} mode an filled by an ideal dielectric:

$$Q_{max} = \frac{0.705R}{\delta} \frac{1 + 2.91R^2/L^2}{1 + 0.86R^2/L^2 + 4.1R^3/L^3} . \qquad (5.20)$$

The loss in stems can be taken into account using the voltage at stems known from (5.17) and current in stems. A detailed analysis of this subject may be found in Avrelin et al. [128]. The final expressions have the form

$$Q = \frac{Q_{max}k_q}{1 + P_s/P_b},$$

$$\frac{P_s}{P_b} = 10.6 \frac{(N-1)l_s R^3 k_p^2 [Z_0\omega_0(C_0 + C_1)]^2}{r_0 L \lambda_0^2 (1 + 0.86R^2/L^2 + 4.1R^3/L^3)}, \qquad (5.21)$$

where P_b reflects the loss in the body of the cavity and $l_s \approx R$ is the stem length. The quality coefficient k_q in these expressions is defined by the manufacturing technology and assembly of the cavity. For a cavity with surface roughness 0.32–0.52 μm assembled by vacuum tight silver brazing, $k_q \approx 0.9$. For stacks of cells (mockups) bolted together, $k_q \approx 0.5$.

The distribution of the electric field in the beam channel is one of the major characteristics affecting the parameters of accelerated particles. It is worth noting that the electric field profile in the form $E_z(z)/(PQ)^{1/2} = F(z)$ is measured experimentally, whereas the quantities R_{sh}, r_{sh}, and U_M are evaluated by integration of this distribution over a certain length. In addition, when one crosses from $U_z(z)$ to R_{sh}/Q, one should roughly estimate $k_g = U/U_M$, where U is the potential difference on the beam channel axis between the centers of adjacent drift tubes and U_M is the potential difference at $r = (a + b)/2$.

The distribution of the electric field in the beam channel is determined by solving the Dirichlet problem for a periodic structure in a quasistatic approximation subject to the given boundary conditions. Using a cylindrical system of coordinates, in the region where particles interact with the field, we have

$$U(r,z) = a_0 + \sum_{k=1}^{k_{max}} a_k I_0\left(\frac{\pi k r}{D}\right)\cos\left(\frac{\pi k z}{D}\right),$$

$$E_z(r,z) = \frac{\pi}{D}\sum_{k=1}^{k_{max}} k a_k I_0\left(\frac{\pi k r}{D}\right)\sin\left(\frac{\pi k z}{D}\right), \qquad (5.22)$$

$$E_r(r,z) = -\frac{\pi}{D}\sum_{k=1}^{k_{max}} k a_k I_1\left(\frac{\pi k r}{D}\right)\cos\left(\frac{\pi k z}{D}\right),$$

where I_0 and I_1 are modified Bessel functions, a_0 and a_k are the coefficients of the expansion, and k_{max} is the number of harmonics taken into account.

The expansion coefficients are calculated by the formulas [132]

$$a_0 = \frac{1}{2M}\left(U_0 + U_M + 2\sum_{i=1}^{M-1}U_i\right),$$

$$a_k = \frac{1}{2MI_0(\pi k R_a/D)}\left(U_0 + (-1)^k U_M + 2\sum_{i=1}^{M-1}U_i\cos\frac{\pi k_i}{M}\right), \qquad (5.23)$$

where R_a is the aperture radius, and U_0, U_1, and U_M are the potentials on a generator of a cylinder of radius R_a and length D taken at $M + 1$ equidistant points.

A model representation in the form of truncated series (5.22) takes into account the cantilever of the electric field and butt-end rounds of drift tubes. It is the most adequate model for the real distribution of the field in the beam channel with cylindrical drift tubes.

Figure 5.8 represents the potential and electric field strength at the axis of accelerating gap of a H-type resonator with a typical geometry of drift tubes calculated by the computer code

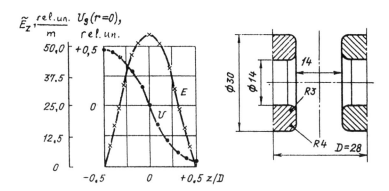

Figure 5.8. Distribution of the potential and electric field strength along the axis of a cylindrical accelerating gap calculated by the MONTEC computer code.

MONTEC-41 [128]. The theoretical distribution coincides with the experimental distribution $E_z(z)$ represented by dots accurate to within +3%. The gap efficiency $k_g = U/U_M = 0.92$. The distribution of potential and, hence k_g, remain constant when all gap dimensions are scaled simultaneously.

For axially symmetric gaps, Figs. 5.9 and 5.10 present plots to determine the efficiency and normalizing factor in a wide range of the parameters l_T/D, l_T/a, and a/D.

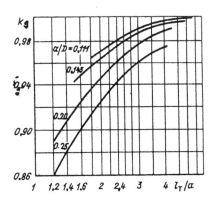

Figure 5.9. To determine the efficiency of cylindrical gaps at $b/a = 2.5$.

Figure 5.10. To determine the normalizing coefficient of the electric field at the center of axially symmetric gaps at $b/a = 2.5$.

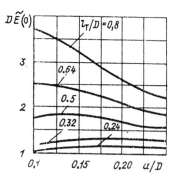

In order to analyze the impedance characteristics of an H-type resonator it is expedient to represent expression (5.19) as the normalized impedance per unit length

$$\lambda_0 \frac{r_{sh}}{Q} = \frac{\lambda_0 R_{sh}}{LQ} = 1.93 \cdot 10^3 \frac{k_p^2 k_g^2}{\beta_{ph}^2 \varepsilon_1}, \qquad (5.24)$$

where $\varepsilon_1 = 2(C_0 + C_1)/\varepsilon_0 D$ is the equivalent permittivity of the dielectric that provides a real retardation. At typical values of $\beta_{ph} = 0.02$, $k_p^2 = 0.41$, $\varepsilon_1 = 15$, $k_g^2 = 0.8$, $Q = 5 \times 10^3$, and $\lambda = 2$ m, $r_{sh} = 260$ MΩ/m. This estimate indicates that, in the region of high retardation, the H-type resonator is a unique structure outperforming other systems in impedance characteristics.

These figures correspond, in the order of magnitude, to the values obtained for known accelerating systems or stakes of cells, however it is expedient to compare the calculated data with results for specific structures. Table 5.2 summarizes experimental and theoretical data for five mentioned H-type cavities with cylindrical drift tubes.

In modern technology of manufacturing and assembly cavities by vacuum tight silver brazing, the quality factor $k_q \approx 0.9$; this fact is proved by data for a molecular ion accelerator in Table 5.2. Cavities brazed by Wood alloy has $k_q = 0.7$, whereas stacks of cells whose stems are bolted to the body are characterized by $0.4 \le k_q \le 0.5$. These values of the quality coefficient have been

Table 5.2 Experimental and calculated impedance characteristics of H-type resonators

	Experimental data				
R, m	0.034	0.19	0.049	0.19	0.25
L_1, m	0.078	0.605	0.18	0.605	1.04
λ, m	0.396	1.74	0.505	2.19	2.76
N	13	19	16	23	31
r_0, m	1	5	1.5	5	6
C_0, pF	0.31	0.8	0.38	1.15	0.95
$U_M(0)/(PQ)^{1/2}$, $\Omega^{1/2}$	12	12.2	10.7	10.6	8.9
R_{sh}/Q, kΩ	10.4	27	14	31	73
Q	2500	5500	3300	4500	9500
	Calculated data				
$U_M(0)/(PQ)^{1/2}$, $\Omega^{1/2}$	11.7	12.3	10.9	10.9	10
R_{sh}/Q, kΩ	10.5	26	15	28	77
Q at $k_q = 1$	3840	11700	4770	10700	11600
Q/k_q	2700	5800	3300	5300	10400
	0.7	0.5	0.7	0.5	0.9

verified in coaxial cavities having an exact analytical solution for the Q-factor and manufactured with the same technology. With allowance for these remarks and data of Table 5.2, formulas (5.16)–(5.21) are suitable for qualitative estimations in the conceptual phase of accelerating systems built around an H-type resonator.

A calculation of an axially symmetric H-type cavity with $\lambda = 2$ m, $L = 1$ m, $a = 5$ mm, $b = 12.5$ mm, $l_T/D = 0.5$, and various values of β_{ph} conducted with the use of graphs and formulas of Sections 5.2 and 5.3, is summarized in Table 5.3. This table includes initial, intermediate, and final calculation results. It also includes the electric field strength parameter in the central gap and the rf power for the given $E(0) = 100$ kV/cm. This table suggests that, for a constant cross section of drift tubes, $r_{sh}(\beta_{ph}) \sim$

Table 5.3 Calculating the body radius and EDCs of an H-type rosonator

β_{ph}	0.05	0.033	0.025	0.020
D_1, mm	50	33.3	25	10
a/D	0.10	0.15	0.20	0.25
C_0, pF	0.860	0.935	1.02	1.10
C_1, pF	0.455	0.314	0.258	0.218
ε_1	6.94	9.45	12.5	15.8
R, m	0.228	0.195	0.168	0.149
r_{sh}/Q, kΩ/m	22.3	36.5	46.5	52.5
$\hat{E}@(0)$, 1/m	34	52.6	66.8	77.5
$E_z(0)/(PQ)^{1/2}$, $\Omega^{1/2}$m	384	517	572	594
k_g	0.995	0.988	0.965	0.922
Q_{max}	2.9×10^4	2.5×10^4	2.17×10^4	1.92×10^4
Q at $k_q = 1$	1.46×10^4	1.28×10^4	1.16×10^4	1.08×10^4
Q at $k_q = 0.7$	1.02×10^4	8.96×10^3	8.12×10^3	7.56×10^4
r_{sh}, MΩ/m at $k_q = 0.7$	227	327	378	398
P, kW at $E(0) = 10$ MV/m	66	42	38	37
P_{st}/P_q at $k_{vd} = 1$	2.4	2.3	2.1	1.9
Q at $k_q = 1$ and $k_{vd} = 1$	6×10^3	5.3×10^3	4.9×10^3	4.63×10^3
r_{sh}, MΩ/m at $k_{vd} = 1$ and $k_q = 0.7$	326	472	556	593

$1/\beta_{ph}$. A part of data in this table refers to the H-type resonator tuned to a uniform distribution of the electric field $E_z(z_n) =$ constant ($k_{vd} = 1$). Let us estimate the effect of the gap capacitance on the impedance characteristics of the H-cavity using the data of the third column. If the outer radius of the drift tube is reduced by 20% to $b = 10$ mm, then $C_0 = 0.78$ pF, $C_1 = 0.23$ pF, $\varepsilon_1 = 10.13$, $R = 0.184$ m, $r_{sh}/Q = 57.5$ kΩ/m, and $Q = 8600$ at $k_q = 0.7$ and $r_{sh} = 495$ MΩ/m. Thus, the shunt impedance has increased by 30% (from 378 to 495 MΩ/m), therefore, the optimization of the dynamic range of H-type cavities with respect to capacitance of accelerating gaps is an important problem of each new ion linac project.

5.4 Axially Asymmetric H-Type Cavities

H-type cavities with an axially asymmetric beam channel may be used to accelerate band ion beams. The typical designs of such cavities are illustrated in Figs. 5.11 and 5.12. Figure 5.11 shows an H-type cavity with a cylindrical body and an accelerating gap made in the form of a grid of rods. Figure 5.12 shows an H-type cavity with a rectangular body and accelerating gaps formed by elongated frames. For an H-type cavity with cylindrical body, the distribution of the magnetic field, adopted in accordance with a chosen cavity analogue, corresponds to system (5.1), therefore to calculate its characteristics one may resort to equations (5.9) and (5.16)–(5.21) with C_0 and C_1 corresponding to a grid-forming gap.

The capacitance of an accelerating gap, made as a grid of rods or frames of rods, may be determined using a relation for the

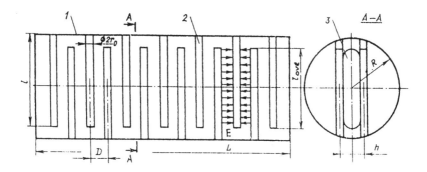

Figure 5.11. H-type cavity with a cylindrical body and accelerating gaps formed by a grid of rods: *1*, body; *2*, rod; and *3*, ion beam channel.

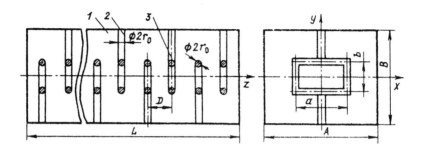

Figure 5.12. H-type cavity with rectangular body and frame accelerating gaps: *1*, body: *2*, frame stem; and *3*, frame.

wave impedance of a symmetric strip line with cylindrical central conductor [133]:

$$C_0 = \frac{\pi \varepsilon_0 / 2}{\ln[2D/(\pi r_0)]} l_{\text{ov}l\Sigma},\qquad(5.25)$$

where $l_{\text{ov}l\Sigma}$ is the total overlapping of the rods forming the gap.

For a grid gap, $l_{\text{ov}l\Sigma} = 2l_{\text{ov}l}$, whereas for a frame gap, $l_{\text{ov}l\Sigma} = 2(a + b)$. The capacitance of rods and frames with respect to the body is estimated, respectively, with the relations

$$C_1^{(\mathrm{r})} \approx \frac{\pi \sqrt{\pi \varepsilon_0} R}{2\ln(R\sqrt{\pi}/2r_0)},$$

$$C_1^{(\mathrm{f})} \approx \frac{4\pi \varepsilon_0 r_0}{\ln\left(\dfrac{A+B}{a+b+2r_0}\right)}. \qquad(5.26)$$

The distribution of the magnetic field in an H-type cavity with rectangular body is approximated by the field of a prismatic cavity resonant in the TE_{101} mode:

$$H_z = B \frac{\lambda_{wg}}{2A} \cos\frac{\pi x}{A} \cos\frac{\pi z}{L},$$

$$H_x = B \frac{\lambda_{wg}}{2L} \cos\frac{\pi x}{A} \sin\frac{\pi z}{L},$$

$$H_y = 0,$$ \hspace{3cm} (5.27)

$$k_z^2 = k_1^2 - k_{cr}^2.$$

Here, the origin of the rectangular coordinate system (see Fig. 5.12) is at the center of the central gap. Recognizing that the critical wavelength $\lambda_{cr} = 2A$, $k_1 = 2\pi\sqrt{\varepsilon_1}/\lambda_0$, and $\lambda_{wg} = 2L$, we derive an expression for the resonance wavelength of an H-type cavity with rectangular body

$$\lambda_0 = 2A \sqrt{\frac{1 + 2(C_0 + C_1)/(\varepsilon_0 D)}{1 + A^2/L^2}} . \hspace{2cm} (5.28)$$

The applicability of relations (5.9), (5.25) and (5.26) is illustrated by Table 5.4 containing experimental and theoretical resonance frequencies of H-type cavities with grid gap and cylindrical body similar to the design diagrammed in Fig. 5.11. For all cavities represented in Table 5.4, $L = 0.45$ m, $R = 0.14$ m, and $r_0 = 4$ mm.

Table 5.4 Testing the relations for the resonant frequency of H-type cavities with axially asymmetric beam channel

	Experiment				Theory	
D, mm	h, mm	f_0, MHz	C_0, pF	C_1, pF	f_0, MHz	
15	16	100	2.08	1.01	103	
30	16	177	1.16	1.01	171	
38	16	201	1.01	1.01	197	
23	16	145	1.40	1.01	143	
23	18	143	1.40	1.01	143	
23	20	141	1.40	1.01	143	

The three last rows of this table indicate that the main source of error in evaluating the resonance frequency is the approximate solution of the electrostatic problem on capacitance of accelerating gaps. For example, formula (5.25) does not take into account the cross capacitance of rods constituting the gap and corresponds to the case of $h/D \gg 1$. Relation (5.25) corresponds to the case $h/D \ll 1$, which also leads to some error in the evaluation of the lateral capacitance of gaps. In general, the approximate theory of H-type cavities with axially asymmetric gaps enables one to determine the transverse dimensions of the body accurate to within 4%.

The impedance characteristics of an H-type cavity with rectangular body are defined by formulas (5.4) and (5.5) with the magnetic field approximated in the form (5.27)

$$
\left.
\begin{aligned}
&\xi_M^\square(z) = \frac{H_z(x,y)}{\sqrt{PQ}} \approx \left[\frac{4\lambda_0}{\pi Z_0 ABL(1+A^2/L^2)}\right]^{1/2} \sin\left(\frac{\pi x}{A}\right)\cos\left(\frac{\pi z}{L}\right), \\[2mm]
&\xi_u^\square(z_n) = \frac{U(z_n)}{\sqrt{PQ}} \approx \left[\frac{Z_0 AB}{\pi L \lambda_0 (1+A^2/L^2)}\right]^{1/2}\cos\left(\frac{\pi Z_n}{L}\right), \\[2mm]
&\frac{R_{sh}^\square}{Q} = 1.92\cdot 10^3\,\frac{ABN^2 k_{vd}^2 k_g^2}{L\lambda_0(1+A^2/L^2)}, \\[2mm]
&Q_{max}^\square \approx \frac{Ak_k}{\delta}\left[\frac{1+A^3/L^3}{1+A^2/L^2}+\frac{1}{2}\frac{A}{B}\right]^{-1}, \\[2mm]
&\frac{R_r}{P_k} \approx \frac{8l_{r,eq}(N-1)A^2 B(Z_0\omega_0 C_0)^2 k_{vd}^2}{\pi\lambda_0^2 r_0 L\left[A^3/L^3 + \dfrac{A}{2B}\left(1+A^2/L^2\right)+1\right]},
\end{aligned}
\right\}
\tag{5.29}
$$

where the equivalent length of rods forming frame gaps is $l_{r,eq} = (2B + a/2 - b/2 - 2r_0)$. For H-type cavities with grid gaps, $l_{r,eq} \approx R$. The coordinate z_n in formula (5.29) corresponds to the center of accelerating gap n. For a detuned uniform H-type cavity, as in a system with axially symmetric gaps, $k_{vd} = 2/\pi = 0.64$. The remarks about the quality coefficient k_q made in Section 5.3 hold also for the present case.

5.5 On-Axis Field in Axially Asymmetric Gaps

Accelerating gaps with large aperture made as a grid composed of cylindrical rods (Fig. 5.11) or frames (Fig. 5.12) can be solved analytically if one uses the solution to the problem of a charged thread between two grounded planes as an approximation [129]:

$$\left. \begin{aligned} U_z(z) &= 2F_0 \tanh^{-1}\left[\frac{\sin(\pi z/D)}{\cosh(\pi h/2D)}\right], \\[2mm] E_z(z) &= F_0 \frac{2\pi}{D} \frac{\cosh(\pi h/2D)\cos(\pi z/D)}{\cosh^2(\pi h/2D) - \sin^2(\pi z/D)}. \end{aligned} \right\} \quad (5.30)$$

In configurations with comparable dimensions r_0, D, and h, the error of the chosen approximation may be compensated by a choice of a point on the rod surface used for normalization of relations (5.30). A qualitative analysis [128] has indicated that the normalization should be performed with point labeled 1 in Fig. 5.13. Then

$$F_0 = \frac{\xi_u \sqrt{PQ}/2}{\tanh^{-1}\left[1/\cosh(\pi r_0/D)\right] + \tanh^{-1}\left\{1/\cosh\left[\pi(h-r_0)/D\right]\right\}} \quad (5.31)$$

and

$$k_g = \frac{\tanh^{-1}\left[1/\cosh(\pi h/2D)\right]}{\tanh^{-1}\left[1/\cosh(\pi r_0/D)\right] + \tanh^{-1}\left\{1/\cosh\left[\pi(h-r_0)/D\right]\right\}}. \quad (5.32)$$

Relations (5.30)–(5.32) provide a sufficient accuracy for practical estimations for $r_0 \geq h/4$ and $r_0 \geq D/4$. This statement is verified by Figs. 5.13 and 5.14 which present experimental and

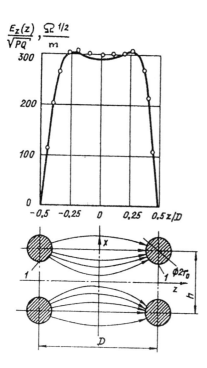

Figure 5.13. Calculated and experimental distributions of an electric field parameter over the axis of a grid gap for $r_0/h = 0.25$ and $h/D = 1.07$.

theoretical (formula 5.30) distributions $E_z(z)/(PQ)^{1/2} = F(z)$ along the axis of a grid gap with densely crowded rods.

For convenience of engineering estimations, expressions (5.31) and (5.32) are plotted in Figs. 5.15 and 5.16. Table 5.5 summarizes

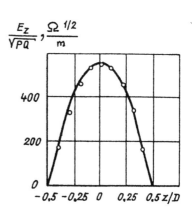

Figure 5.14. Calculated and experimental distributions of an electric field parameter over the axis of a grid gap for $r_0/h = 0.25$ and $h/D = 0.55$.

Figure 5.15. Efficiency coefficient of a grid or frame-type gaps.

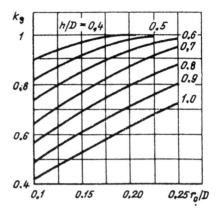

calculated and experimental data for six H-type mockups (similar to the one depicted in Fig. 5.11). In these stacks of cells, accelerating gaps were made as rod grids. For all mockups, $l_{ov|\Sigma}$ = 0.13 m, R = 0.14 m, L = 0.45 m, r_0 = 4 mm, and $k_{vd} \approx 1$. The stacks of cells in mockups were fixed by bolting.

The distributions of the electrical field along the axis of the beam channel in mockups 1 and 3 of Table 5.5 correspond to Figs. 5.13 and 5.14. This table reveals a considerably lower efficiency of rod and frame gaps compared to tube gaps. For the distribution of the longitudinal electric field over the cross-section

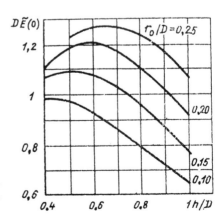

Figure 5.16. Plots to determine the normalizing factor of the electric field strength at the center of grid or frame gaps.

Table 5.5 Experimental and calculated characteristics of H-type cavity with grid gap

	Mockups					
	(1)	(2)	(3)	(4)	(5)	(6)
	Inputs					
D, mm	15	30	38	23	23	23
h, mm	16	16	16	16	18	20
N	23	13	11	15	15	15
β_{ph}	0.010	0.036	0.051	0.022	0.022	0.022
f_0, MHz	100	177	201	145	143	141
$\xi_e(0)$, $\Omega^{1/2}$/m	555	390	307	480	410	390
	Experiment					
r_{sh}/Q, kΩ/m	39	32	36	31	26	22
Q	3200	2700	2600	2900	3000	3100
	Calculation					
f_0, MHz	103	171	197	143	143	143
$\xi_e(0)$, $\Omega^{1/2}$/m	550	376	294	462	437	402
r_{sh}/Q, kΩ/m	37	31	26	30	25	21
Q at $k_q = 0.5$	3500	2700	2500	3100	3100	3100
k_g	0.7	0.86	0.88	0.81	0.75	0.68

of a grid and frame gaps (for $a \gg b$ or $b \gg a$) at $z = 0$, the inhomogeneity can be calculated by the formula

$$\frac{E_z(x)}{E_z(x=0)} = \frac{\cosh[\pi h/(2D)]}{2}\left\{\frac{1}{\cosh(\pi x/D)} + \frac{1}{\cosh[\pi(h-x)/D]}\right\}. \tag{5.33}$$

For mockup 5 of Table 5.5, it is presented in Fig. 5.17 in the form $E_z(\Delta)/E_z(x=0) = F(\Delta)$, where $\Delta = x/(h/2 - r_0)$ is the measure of deviation of a particle from the cavity axis. This figure demonstrates that the accelerating gaps compiled of rods provide for the distribution nonuniformity $\delta E/E = [E_z(0) - E_z(\Delta)]/E_z(0) \leq 5\%$ in the range $-0.8 \leq \Delta \leq 0.8$.

Figure 5.17. Nonuniformity of accelerating electric field in the middle plane of a grid or frame gaps for $h/D = 0.8$ and $r_0/h = 0.22$.

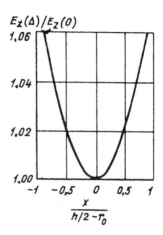

The overvoltage coefficient of H-type cavities is defined as $k_{ov} = E_M/E_z(z_n)$, where z_n is the coordinate of the center of a gap. The overvoltage coefficient of accelerating gaps formed by rod grid or elongated frames (Figs. 5.11 and 5.12) can be estimated using the electrostatic analogy

$$k_{ov} \approx \frac{1}{\tilde{E}D} \frac{\pi}{2\cos\left[\dfrac{\pi}{2}\left(1 - \dfrac{2r_0}{D}\right)\right]\tanh^{-1}\left\{\sin\left[\dfrac{\pi}{2}\left(1 - \dfrac{2r_0}{D}\right)\right]\right\}}. \qquad (5.34)$$

The maximum field strength is found on a straight line parallel to the rod axis at a distance from the cavity axis exceeding $h/2$ by $(0.1 \pm 0.03)r_0$. Equation (5.17) is worked out in Fig. 5.18.

In axially symmetric accelerating gaps (Fig. 5.1), for $R_{r1} > R_{r2}$ the highest electric field strength is achieved on the surface of drift tube rounds. Figure 5.19 plots experimental values of k_{ov} obtained with a ring probe in a mockup of an H-type cavity with $\lambda_0 = 2$ m, $R_{r1} = R_{r2} = 2.5$ mm, $a = 5$ mm, and $l_T = 15$ mm. Measurement error is within 5%. As follows from this figure, H-type cavities with axially symmetric beam channel are characterized by $1.5 \leq k_{ov} \leq 2.4$.

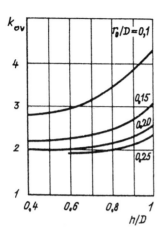

Figure 5.18. Overvoltage coefficient of grid or frame accelerating gaps.

The results of this section lead us to conclude that the approximate method of analysis of H-type cavities with axially asymmetrical beam channel is a reliable approach for express analysis of the dynamics of this type of slow wave systems. This method may be useful at the conceptual phase of H-type cavity development, and on the stage of scheduling experiments associated with the tuning of H-type cavities to a given distribution of the field in the gaps.

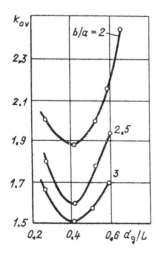

Figure 5.19. Overvoltage coefficient of axially symmetric accelerating gaps at $R_r = (b - a)/3$. Open circles represent experimental values.

5.6 Tuneup to a Given Electric Field Profile in Gaps

As a rule, the distribution of voltage between accelerating gaps corresponding to an optimal particle dynamics is closer to uniform than to a cosine law. One can tune H-type cavities to a given distribution of $U_M(z)$, for example, with additional circuits connected to gaps. The idea behind this method is highlighted in Fig. 5.20. In the initial position, the voltage at the end gap is determined by the flow of induction $\mu_0 H_\varphi S$ through the contour described by points 1, 2, 3, and 4. If the stem of the end drift tube is made as a rod with configuration 2, 5, 6, and 7, then the voltage U_I is defined by the flux of inductance through the contour 7, 6, 5, 2, 3, 4, and 7. For the purpose of analysis we resort to equations (5.1) and (5.4) to define U_I and U_{II} and the voltage at the central gap U_M as follows

$$
U_I = \kappa \left\{ \sin \frac{\pi D}{L} \int_0^{\mu_{11}} \frac{J_1(x)}{x} dx \right.
$$

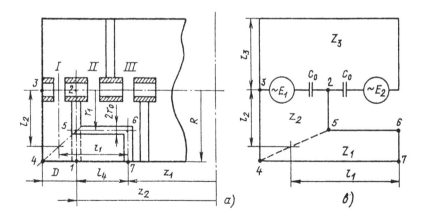

Figure 5.20. (a) Physical and (b) equivalent circuits of a stem leveling the voltages at end accelerating gaps.

$$+ 2 \cos \left[\frac{\pi}{L} \left(D + \frac{l_4}{2L} \right) \right] \sin \frac{\pi d_4}{2L} \int_{\mu_{11} r_1 / R}^{\mu_{11}} \frac{J_1(x)}{x} dx \Bigg\},$$

$$\left. \vphantom{} \right\} \quad (5.35)$$

$$U_{\mathrm{II}} = U_{\mathrm{I}} + \kappa \sin \frac{2\pi D}{L} \int_0^{\mu_{11}} \frac{J_1(x)}{x} dx,$$

$$U_M = 2\kappa \int_0^{\mu_{11}} \frac{J_1(x)}{x} dx,$$

where $\kappa = B\mu_0 R^2 \omega / \mu_{11}$, l_4 is the length of the horizontal part of the stem indicated in Fig. 5.20, and r_1 is the distance of this part from the axis of the structure. For $x < 2$, the integrals in these expressions are computed as the series

$$\int_x^{\mu_{11}} \frac{J_1(x)}{x} dx \approx 0.801 - \frac{x}{2} + \frac{1}{6} \left(\frac{x}{2} \right)^3 - \frac{1}{60} \left(\frac{x}{2} \right)^5. \qquad (5.36)$$

Let, for example, $L = 0.60$ m, $D = 2.5 \times 10^{-2}$ m, and $r_1/R = 0.4$. Then, in the initial state with $l_4 = 0$, we will have $U_{\mathrm{I}} : U_{\mathrm{II}} : U_M = 0.065 : 0.194 : 1$. Thus, a longer horizontal section of the stem, l_4, levels the voltage with respect to the central gap. At some length l_4, the resonant frequency of the structure including two end gaps, their stems and the body coincides with the resonant frequency of the H-type cavity. In this case. in calculating the relation $U_{\mathrm{I}} : U_{\mathrm{II}} : U_M$, one should take into account resonant phenomena.

The equivalent circuit of this system may be constructed from sections of strip lines formed by rods and the body and lumped capacitances of the gaps C_0. As seen in Fig. 5.20b, the system is excited by the emfs $E_1 = U_{\mathrm{I}} \sin \omega_0 t$ and $E_2 = U_{\mathrm{II}} \sin \omega_0 t$. The wave impedance Z_1 corresponds to an asymmetric coaxial line, Z_2 corresponds to an asymmetric strip line, and Z_3 corresponds to a symmetric strip line. To determine the resonance frequency of the system we use the known rule $\Sigma_i \mathrm{Im} Z_i = 0$, where Z_i is the impedance on the right and on the left from a chosen cross section. For section *3–2* in Fig. 5.20, we have

$$\frac{\tan \beta l_1 + m \tan \beta l_2}{1 - \dfrac{1}{m} \tan \beta l_1 \tan \beta l_2} + \frac{Z_3 \tan \beta l_3 - 1/(\omega_c C_0)}{2 - Z_3 \omega_c C_0 \tan \beta l_3} = 0, \tag{5.37}$$

where $\omega_c = 2\pi c/\lambda_c = \beta c$ is the resonance circular frequency of the circuit, $m = Z_2/Z_1$, and $c = 3 \times 10^8$ m/s. For a nonsymmetric coaxial line,

$$Z_1 = \frac{Z_0}{2\pi} \tanh^{-1} \frac{R^2 + r_0^2 - r_1^2}{2R r_0}. \tag{5.38}$$

Estimations of Z_1 and Z_2 with any model of strip lines [134] indicate that for the ratios of dimensions R, r_0, and r_1 typical of H-type cavities, $Z_1 \approx Z_2 \approx Z_3 = 200\ \Omega$. Consequently, for rough estimation, equation (5.37) may be simplified:

$$\tan\left[\beta(l_1 + l_2)\right] + \frac{\tan \beta l_3 - 1/(\omega_c C_0 Z_1)}{2 - Z_1 \omega_c C_0 \tan \beta l_3} = 0. \tag{5.39}$$

Equations (5.37) and (5.39) assume that the length of lines constituting the oscillating system are far larger than the length of segments containing capacitance gaps and length of a short circuit line between points *6* and *7* in Fig. 5.20. In our case, the length of these segments are comparable and equation (5.39) should be treated as the lower bound for the resonance wavelength λ_c.

We used equation (5.39) to estimate the length of the horizontal part of the stem $l_4 = l_1 - D/2$ for $\lambda_c = 2.0$ m, $R = 0.19$ m, $D = 2.5 \times 10^{-2}$ m, $r_1/R = 0.4$, $r_0 = 5 \times 10^{-3}$ m, $l_2 = (R + r_1)/2$, $l_g = R$, and $C_0 = 10^{-12}$ F. We obtained $Z_1 = 210\ \Omega$, $l_1 + l_2 = 0.371$ m, and $l_4 \le 0.225$ m. This length of the horizontal segment of the stem fits into the design dimensions of a typical H-type cavity with the operating wavelength $l_0 = 2$ m. Consequently one can take advantage of using the tuning system in the resonance mode.

For a circuit containing a capacitor loaded onto a x_L-r_a series circuit and an alternating emf with amplitude U_I, the amplitude of the voltage on the capacitor is

$$U_I' = U_I \left\{ \left[1 - \left(\frac{\lambda_c}{\lambda_0} \right)^2 \right]^2 + \frac{1}{Q_c^2} \left(\frac{\lambda_0}{\lambda_c} \right)^2 \right\}^{-1/2}, \qquad (5.40)$$

where $Q_c = \omega_c L/r_a$ is the circuit quality factor, L is the circuit inductance, and λ_0 is the wavelength of the alternating emf. For $\lambda_c \ll \lambda_0$, equation (5.40) indicates that the voltage across the capacitance is equal to the amplitude of the source. For $\lambda_c = 2$ m, one can estimate the quality factor of the system represented in Fig. 5.20 using available reference data [133] as $500 \leq Q_c \leq 1500$. For $U_I' \leq 20 U_I$, even the lowest Q_c leads to

$$U_I' \approx U_I \left[1 - \left(\frac{\lambda_c}{\lambda_0} \right)^2 \right]^{-1}. \qquad (5.41)$$

Using the last formulas and system (5.35), we rewrite the ratio of voltages of central and end gaps as

$$\frac{U_c}{U_I} \approx \frac{0.5 \sin \left(\frac{\pi D}{L} \right) + 1.25 \cos \left[\frac{\pi}{L} \left(D + \frac{l_4}{2} \right) \right] \sin \left(\frac{\pi l_4}{2L} \right) \int_{\mu_{11} r/R}^{\mu_{11}} \frac{J_1(x)}{x} dx}{1 - \left(\frac{\lambda_0}{\lambda_c} \right)^2},$$

$$(5.42)$$

where l_4 is related to λ_c by the transcendental equation (5.39). A similar expression can be derived also for U_M/U_{II}. Results of the simultaneous solution of (5.39) and (5.42) for the inputs of the preceding example are summarized in Table 5.6.

Thus, voltage across the central and end gaps has been leveled at $l_4 \approx 0.16$ m, and the voltage across the central and the second gap from the end has been leveled at $l_4 \approx 0.13$ m.

Table 5.6 Leveling the voltage at gaps

l_4, m	λ_c, m	U_I/U_M	U_{II}/U_M	U_{II}/U_I	U_{III}/U_M
0.123	1.60	0.61	0.95	1.57	0.32
0.149	1.70	0.89	1.35	1.52	0.32
0.175	1.80	1.4	2.0	1.48	0.32

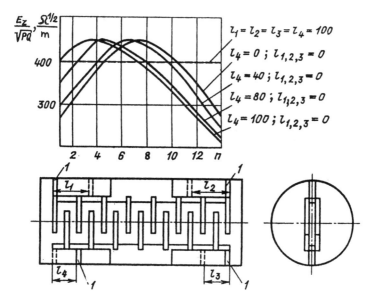

Figure 5.21. Leveling the electric field strength at the center of grid accelerating gaps by end elements of a tuning facility.

A more flexible control of voltage at gaps is achieved by loading onto the in-system circuit all the drift tubes which have identical sign of potential and whose stems fall onto the horizontal part of the longitudinal tie l_4. One can achieve a more flexible tuning configuration with the use of several tuning facilities. Figure 5.21 shows an H-type cavity corresponding to mockup 4 in Table 5.5. This design is equipped with four tuning facilities covering four end gaps each.

We tuned the cavity by moving short-circuiting elements *1*. When one of these short-circuiting elements was allowed to move, the amplitude envelope at the centers of gaps $E_z(n)(PQ)^{1/2}$ ~ $U_M(n)$, where n is the gap number that varies as indicated in the diagram. In agreement with the previous qualitative estimates, in end gaps, the parameter $E_z(n)/(PQ)^{1/2}$ increases with l_4, and, at $l_4 \approx 0.1$ m, the voltage at end gaps exceeds the voltage at the central gap. A uniform distribution $E_z(n)/(PQ)^{1/2} =$ constant is attained when all tuning facilities are engaged. Operation of these tuning facilities can be analyzed by compiling an equivalent circuit, similar to that shown in Fig. 5.20b, and deriving an equation of the type (5.39) for the resonance wavelength of this circuit.

An exact quantitative theory of tuning facilities must take into account the coefficient of mutual inductance M between H-type cavities and installed circuits. Estimations indicate that $0.1 \leq M/L_{eq} \leq 0.2$. A solution to this self-consistent problem allows one to determine the variation of resonance frequency, reduction of the Q-factor and voltage at the central gap, since a proportion of the main magnetic flux is intercepted by tuning facilities. The last element has been inserted in formulas (5.19) and (5.29) as a distribution coefficient $k_{vd} \leq 1$. For the case indicated in Fig. 5.21, $k_{vd} \approx 0.9$.

A system with four tuning facilities covering end accelerating gaps allows one to obtain an essentially nonuniform distribution of $E_z(n)/(PQ)^{1/2}$ corresponding to an ion beam with sign-reversing rf focusing. One version of such a tuneup of an H-type cavity has been analyzed by Zverev et al.[134].

6

MEASUREMENT OF ACCELERATING CAVITIES

6.1 General

Experimental electrodynamic studies of accelerating structures are an important stage in the development of any accelerator of charged particles. The need for such studies stems from an insufficient accuracy of analytical and numerical techniques used to design accelerating cavities and from the desire to perfect the cavity technology. In this connection, one should distinguish two stages in electrodynamic measurements of accelerating cavities. In the first stage, mockup measurements are carried out to determine the necessary electrodynamic characteristics (EDCs) as functions of geometric dimensions, to test different methods of tuning and correcting the electric field and resonance frequency in accelerating gaps, to examine the characteristics at higher order modes (HOMs), and so on. In the second stage, the EDCs of manufactured accelerating cavities are tested and tuned, if necessary.

Experience in the field of metrology of accelerating structures has been gained by many accelerator centers. A comprehensive analysis of publications on this subject reveals a great difference in studied parameters, methods, and structures.

All the methods of experimental studies of accelerating structures can be conventionally divided into the methods for standing-wave structures and for traveling-wave structures. Detailed data on measurements in TW structures can be found in our handbook [3], which contains an extensive literature on this subject and unique procedures for measuring the electrodynamic characteristics and for tuning DLWs.

The resonance method undoubtedly plays a leading part in experimental studies of accelerating cavities. The advantages of this method are as follows:

(1) accuracy of measurements and their information content,

(2) amenability to automatic measurements,

(3) a range and high quality mass produced instrumentation for measuring signal frequencies and amplitudes,

(4) a one-to-one correspondence between the characteristics of a uniform periodic structure and its resonance mockup, and

(5) a relatively simple processing of measurement data.

Because of these advantages, the resonance method is indispensable in large-scale electrodynamic research of accelerating structures.

Direct measurements with single cells determine the resonance frequencies, and measurements with stacks of coupled cells (mockups) yield dispersion curves in the fundamental and higher order modes. Other parameters are determined by indirect measurements from the frequency characteristics. These measurements include

(1) the Q-factor or the transmission coefficient of cavities,

(2) the electric field strength in the beam channel and in places critical for sparking, and

(3) the profile of the electromagnetic field to identify cavity modes and calculate the distribution of losses in the cavity walls.

One also must perform measurements to equalize the accelerating field, especially in long (many-cell) cavities, tune fabricated accelerating cells to the specified EDCs, tune rf couplers to the given coupling coefficient; and symmetrize the couplers.

A significant part of these measurements can be conducted by the bead-pulling technique using semi-automated measuring systems [135, 136]. This chapter is mainly devoted to the development of the bead-pulling technique and to its use in measurements of EDCs of accelerating cavities. The measurement techniques and circuits developed for this purpose will be demonstrated using BPSs and interdigital H-type cavities as examples.

6.2 Instrumentation

Figure 6.1 shows the block diagram of a device used to measure resonance frequencies, Q-factors, and transition loss of accelerating cavities. This test bench comprises usual instruments for the respective frequency band.

The cavity under study is connected by a through circuit using two symmetric couplers. Given an output power of rf generators above 10^{-2} W and a minimum sensitivity of the spectrum analyzer

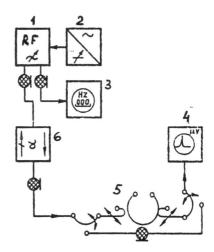

Figure 6.1. Block diagram of the. test bench: *1*, rf generator; *2*, power supply, *3*, frequency counter; *4*, spectrum analyzer; *5*, cavity with couplers; and *6*, attenuator

about 10^{-8} W per mm of response height on the scope, the transition loss of the cavity A can be greater than 40 dB. Resonance corresponds to the maximum amplitude of the response on the spectrum analyzer scope (Fig. 6.2).

The maximum error in the measured resonance frequency of the cavity is

$$\frac{\Delta f_0}{f_0} = \frac{\Delta f_e}{f_0} + \frac{\Delta f_g}{f_0} + \frac{\Delta f_f}{f_0} + \frac{\alpha \Delta T}{f_0}, \tag{6.1}$$

where Δf_e is the error in reading the response position, Δf_g is the frequency instability of the generator during measurements, Δf_f is the error of the frequency meter, and ΔT is the temperature drift during measurement.

Usual instrumentation of the bench ensures the relationship $(\Delta f_g + \Delta f_f + \alpha \Delta T) \ll \Delta f_e$. The quantity $\Delta f_0/f_0$ is calculated from the equation of the cavity Q-factor in the square detection law

$$P(f) = \left[1 + \left(2Q_1 \frac{f_g - f_0}{f_0} \right)^2 \right]^{-1} P_0 \tag{6.2}$$

and has the form

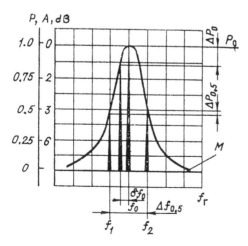

Figure 6.2. Response on the screen of a spectrum analyzer.

$$\frac{\Delta f_0}{f_0} = \frac{1}{2Q_1} \sqrt{\frac{\Delta P_0}{P_0}} \, , \tag{6.3}$$

where ΔP_0 is the error in the position of the response about the amplitude value P_0 shown in Fig. 6.2. We find that $\Delta f_0/f_0 \leq 0.07\,Q_1^{-1}$ at response height $P_0 \approx 100$ mm and at $\Delta P_0 \leq 2$ mm.

In deriving the dispersion equation of the structure, the resonance frequency (or wavelength) is put in correspondence to the dimensions of the accelerating structure. It is obvious that in this case, errors in measured data must be determined with an eye to the tolerances for the dimensions of the cells. In practice, the cells have at least five dimensions with independent tolerances. The deviation of the resonant frequency of a resonance mockup associated with the tolerances for dimensions q_i is defined as

$$\frac{\Delta f_0}{f_0} = \frac{1}{f_0} \sqrt{\sum_i \left(\frac{\partial f}{\partial q_i} \Delta q_i \right)^2} \, . \tag{6.4}$$

For a DLW with rms dimensional tolerances $\Delta q_i = 0.01$ mm, the error $\Delta f_0/f_0$ calculated by formula (6.4) is an order of magnitude greater than the random error due to instruments of the test bench.

Relations (6.3) and (6.4) serve as a basis for the estimation of random errors in phase and group velocities, the coefficient of coupling between cells, and other parameters related to the measured resonance frequencies [3, 122, 137].

The loaded Q-factor of a cavity is defined as $Q_1 = f_0/(f_2 - f_1)$, where f_2 and f_1 are the half-power frequencies. The response positions corresponding to the frequencies f_1 and f_2 are shown in Fig. 6.2. The rms error in Q_1 measurements is

$$\frac{\Delta Q_1}{Q_1} = \frac{2\Delta P_{0.5}}{P_{0.5}} + \frac{\Delta f_g}{\Delta f_{0.5}} + \frac{\Delta f_f}{\Delta f_{0.5}} + \frac{\alpha \Delta T}{\Delta f_{0.5}} + \frac{\Delta P_g}{P_g} + \frac{\Delta f_0}{f_0} \, , \tag{6.5}$$

where $\Delta f_{0.5} = f_2 - f_1$ and ΔP_g is the instability in the generator output power within the band of retuning $\Delta f_{0.5}$. Here, as in (6.3), the rms error $\Delta P_{0.5}$ at the position of response is significantly greater than the other terms, which enables us to replace (6.5) by the relation $\Delta Q_l/Q_l \leq 2\Delta P_{0.5}/P_{0.5}$. For the half response height, we have $\Delta P_{0.5}/P_{0.5} \leq 2\cdot 10^{-2}$ and $\Delta Q_l/Q_l \leq 4\cdot 10^{-2}$.

The loaded and unloaded Q-factors of the cavity are related by

$$Q_0 = Q_1\left(1 + 10^{-A/20}\right), \tag{6.6}$$

where A is the transition loss of the cavity. The transition loss is measured on the bench by the replacement method using calibrated attenuators of the spectrum analyzer. According to the setup certificate, the absolute error of such measurements is $|\Delta A| \leq 1$ dB. At $A = 30$ dB, we have $Q_1 = 0.97 Q_0$, so that the effect of a coupler may be disregarded in a number of cases.

The unloaded Q-factor can be determined if the effect of short-circuiting plates forming the resonance mockup is ignored. This is equivalent to measurements of the Q-factor of a mockup with length $L \to \infty$. However, Q_0 can be determined from measurements of the unloaded Q-factors of two resonance mockups with different lengths. In this case [138],

$$Q_\infty = \frac{Q_1 Q_2 (L_2/L_1 - 1)}{L_2 Q_1/L_1 - Q_2}, \tag{6.7}$$

where Q_1 and Q_2 relate to mockups with lengths L_1 and L_2, respectively. It is sufficient to use $L_1 = \lambda_w/2$ and $L_2 = \lambda_w$, so that (6.7) will yield $Q_\infty = Q_1 Q_2 (2Q_1 - Q_2)^{-1}$.

Measurements of the resonance-frequency and Q-factor are accompanied by systematic errors due to temperature drift, coupler effects, and mockup deformation. The first of these errors is eliminated by reducing the measurements to the room temperature $T = 20°C$ by the formula $f_0(T_{st}) = f_0(T)[1 + \alpha(T_{st} - T)]$, where T_{st} is the temperature of measurements and, for copper, $\alpha = 1.65\cdot 10^{-5}$ K^{-1}.

Experimental curves describing the effect of coupling loops are derived in the form $f_0 = F(l_1, h_1, r_0)$, where l_1 is the depth to which the loop is lowered into the cavity, h_1 is the width of the loop, and r_0 is the radius of a wire forming the loop. An analysis of these curves leads us to the conclusion that, at a transition loss $A \geq 25$ dB, the influence of a coupler on the resonant frequency of the mockup is estimated at less than $2 \cdot 10^{-5} f_0$ and may be neglected.

The resonance mockups of DLWs, BPSs, H-type resonators, and other studied structures are formed as stacks of cells or a finite number of modules, for example, rings and irises. The electrical contact between cells is formed under a load of a press. It was found that the Q-factors of such mockups virtually do not change at specific pressure on the contact surfaces $\sigma = 20$–40 N/mm². In designing new modules, one should avoid configurations with bending moments. All binding stresses must be taken up by the body as compression loads. In this case, the relative change in the longitudinal size of a cell reduces to $\Delta D/D \leq \sigma/E$, where E is the Young modulus of the first kind. For copper, $E \approx 1.1 \cdot 10^5$ N/mm² and $\Delta D/D \leq 3.6 \cdot 10^{-4}$.

Now, we will describe instruments necessary to measure electromagnetic fields by the bead-pulling technique. This method is based on the theorem of small perturbations [11]:

$$\frac{\Delta f_0}{f_0} = \frac{f_1 - f_0}{f_0} = -\frac{1}{W} \sum_{i=x,y,z} \left(\mathcal{H}_i^{(e)} E_i^2 - \mathcal{H}_i^{(m)} H_i^2 \right), \qquad (6.8)$$

where f_1 is the resonance frequency of the cavity perturbed by the probe; the x, y, and z axes are aligned with the probe axes; and $\mathcal{H}_i^{(e)}$ and $\mathcal{H}_i^{(m)}$ are the electric and magnetic probe formfactors along the corresponding axes. If the probe is located in the region of a one-component field, for example, on the axis of a cavity excited in the TM_{010} mode, relation (6.8) will take the form

$$\Delta f_0 / f_0 = -\mathcal{H}_z^{(e)} E_z^2 / W .$$

Using the relation $W = QP/(2\pi f_0)$, we obtain for the parameter of the electric field strength on the cavity axis

$$\xi_e(z) = \frac{E_z(z)}{\sqrt{PQ}} = \sqrt{\frac{\Delta f_0(z)}{2\pi \mathcal{H}_z^{(e)} f_0^2}} \; . \tag{6.9}$$

According to (6.9), to realize the method of small perturbations, it is necessary to have a setup that is capable of measuring frequency deviations of the cavity on the order of $\Delta f_0/f_0 \approx (3\text{--}5) \cdot 10^{-5}$ accurate to within 5% or better.

Figure 6.3a presents the block diagram of a wide-band cavity tester (WBCT) [135, 136] whose operation is based on the bead-pulling technique.

The diagram is divided into three blocks: an amplifier–oscillator section (I), a probe pulling system (II), and a data output system (III). The first block contains a microwave amplifier built around transistors, or a traveling-wave tube with a stabilized power supply of the amplifier. A feedback loop includes a calibrated phase shifter, a ferrite isolator, a cavity filter, and the cavity itself. The phase shifter and controllable couplers with the cavity ensure the phase and amplitude conditions for self-oscillations. The cavity filter ($Q_1 \approx 400$) bounds the passband of the feedback loop within $f_0 \pm 5$ MHz, thus facilitating the excitation of the given mode in the cavity. The travel unit moves the probe within the cavity and forms pulses that trigger the frequency meter in a given interval of probe movement Δz ($0.2 \le \Delta z \le 5$ mm). The potentiometer bound to a driving roller of a probe suspension generates a voltage which is directly proportional to the probe coordinate. This voltage is used to synchronize the motion of the plotter coordinate carriage with the probe movement.

WBCTs operate on the principle of a tracking self-oscillator controlled in frequency by the cavity. The conditions for the existence of self-oscillations are governed by the set of equations

$$\left. \begin{array}{l} k_a \ge \displaystyle\sum_i A_i, \\[2em] \varphi_c + \varphi_a + \varphi_{fb} + \varphi_f = 2\pi n, \end{array} \right\} \tag{6.10}$$

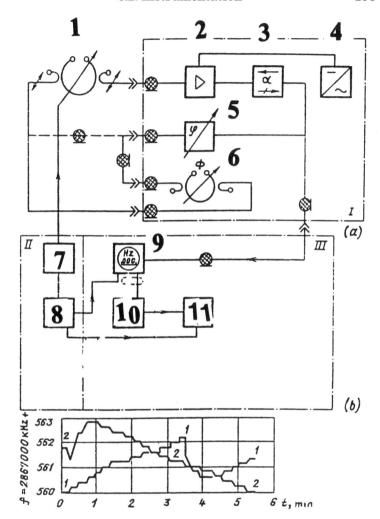

Figure 6.3. Semi-automatic WBCT-type test bench. (a)
Block diagram: *1,* accelerating cavity; *2,* amplifier; *3,* ferrite
isolator; *4,* power supply; *5,* phase shifter, *6,* resonator-filter;
7, probe suspension device; *8,* automatic controller of probe
movement; and *9,* frequency counter; *10,* DAC; *11,* voltmeter.
A computerized test jig is shown by dashed lines. (b) Time
dependence of the oscillation frequency of the WBCT-1 for
the cavity Q-factor $Q_1 = 5 \cdot 10^3$: *1,* after five minute warmup;
and *2,* after one hour warmup. The frequency sampling
interval is $\Delta t = 4$ s.

where k_a is the power gain, A_i is the transition loss in the ith feedback element, φ_c and φ_f are the phase increments in the cavity and its filter, respectively; and $\varphi_{fb} + \varphi_a = \varphi_{0t}$ is the total phase advance in feedback cables and amplifier circuits. A theory of such measuring systems is based on an analysis of system (6.10).

Summarizing the results of this analysis yields the main characteristics of WBCTs [136] as follows:

the short-term instability of self-oscillation frequency

$$\frac{\delta f_0}{f_0} \le \frac{1}{2Q_1} \frac{\partial \varphi_a}{\partial u_0} \Delta U_0 , \tag{6.11}$$

the phase band of locking into a self-oscillation regime

$$\cos(\Delta\varphi_{lock}) \cos\left(\Delta\varphi_{lock} \frac{Q_f}{Q_1}\right) = 10^{-\Delta k_a/20} , \tag{6.12}$$

the maximum deviation of the cavity's resonant frequency

$$\cos\left[k_{ot}\Delta f_0^{(max)} + \arctan\left(\frac{2Q_f}{f_0} \Delta f_0^{(max)}\right)\right] \cos\left(\frac{2Q_f}{f_0} \Delta f_0^{(max)}\right) = 10^{-\Delta k_a/20} , \tag{6.13}$$

and the systematic error in tracking the deviation of this resonant frequency

$$k_s = \frac{\Delta f_0'}{\Delta f_0} \approx \left[1 + \frac{Q_f}{Q_1} + \frac{k_{ot}f_0}{2Q_1}\right]^{-1} . \tag{6.14}$$

Here, Q_1 is the loaded Q-factor of the cavity, Q_f is the Q-factor of the cavity filter, U_0 and ΔU_0 are the supply voltage and its instability, respectively;

$$\Delta k_a = k_a - \sum_i A_i$$

is the gain margin of the setup; and $k_{ot} = \partial\varphi_{ot}/\partial f$ is the slope of the feedback phase–frequency characteristic.

The operating characteristics of setups include the measurement speed S = $v_p/\Delta z$, which is equal to the number of measured points per second, and the measurement capability C = $60v_p$, which is equal to the length of a resonance mockup covered per minute. Here, v_p is the speed of the probe and Δz is the measurement step.

Relations (6.11) and (6.14) and the transcendental equations (6.12) and (6.13) constitute a system for estimating the characteristics of measuring systems and evaluating their metrological potentials.

If the amplifier–oscillator is built around a traveling-wave tube, relation (6.11) takes the form

$$\frac{\Delta f_0}{f_0} = \frac{0.29\pi}{Q_1}\left(\frac{2eU_0}{m_e c^2}\right)^{-1/2}\frac{\Delta U_0}{U_0},\qquad(6.15)$$

where we use the relation for $\partial\varphi_a/\partial U_0$ from [139]. Here, $e/m_e = 1.76\cdot10^{11}$ C/kg and $\Delta U_0/U_0 = 5\cdot10^{-4}$ according to the power supply certificate. For typical values $U_0 = 10^3$ V and $Q_1 = 10^4$, we have $|\Delta f_0/f_0| \leq 7.3\cdot10^{-7}$. The absolute frequency instability $|\Delta f_0| \leq 2.2$ kHz at $f_0 = 3$ GHz is admissible to realize the method of small perturbations.

Further calculations will be performed for the following realistic parameters: $l_{fb} = 4$ m, $l_s = 0.15$ m, $Q_f = 500$, and $\Delta k_a = 3$ dB, where l_{fb} is the length of feedback cables and l_s is the length of the traveling-wave tube spiral. In the traveling-wave tube, the condition $\beta_p \approx \beta_e = (2eU_0/m_e c^2)^{1/2}$ is satisfied; it determines the slope of the feedback phase–frequency characteristic

$$k_{fb} = \frac{d\varphi_{fb}}{df} = \frac{2\pi}{c}\left[l_{fb} + l_s\left(\frac{2eU_0}{m_e c^2}\right)^{-1/2}\right].\qquad(6.16)$$

From (6.16) it follows that $k_{fb} = 0.134$ rad/MHz. Using the above data, we find that $\pm\Delta\varphi = 0.784$ rad = 45°, $\pm\Delta f_0^{(max)} = 1.42$ MHz, and $k_s = 0.93$.

The phase band of locking indicates that the system can be easily tuned to the operating mode: self-oscillations occur in an interval of about one fourth of the phase-shifter operating range. Knowing $\Delta f_0^{(\max)}$ and using (6.9), we can determine the limiting value of the probe formfactor and the minimum random error in the measurements of the electric field parameter $\delta \xi_e / \xi_e \approx \delta f_0 / 2 \Delta f_0^{(\max)} = 0.78 \cdot 10^{-3}$. The value $k_s = 0.93$ characterizes the systematic error in ξ_e equal to $(\delta \xi_e / \xi_e)^{\mathrm{syst}} = -3.6\%$.

The reliability of the above estimates is proved by Fig. 6.3b, which presents the time dependence of the frequency of self-oscillations in a WBCT-1 operating with a cavity having $Q = 9.5 \cdot 10^3$ and placed in a room without artificial micro-climate ($\Delta T = 1°\mathrm{C/h}$). The 5-minute frequency instability is less than $\delta f_0 / f_0 = 5 \cdot 10^{-7}$, which is close to the value estimated by relation (6.15).

When equipped with a jig shown in Fig. 6.4, a general-purpose measuring bench made by Fig. 6.3a can also be used for rapid analysis of cavity shunt impedance.

This device is made as a ring formed by two dielectric threads which differ in thickness by at least a factor of 4–5. The length of each thread is longer than the length of the cavity. The

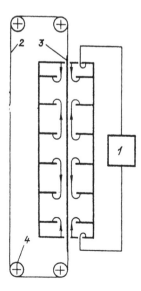

Figure 6.4. Jig for rapid measurements of cavity shunt impedances: *1*, general-purpose test bench; *2*, thin dielectric thread; *3*, thick dielectric thread; and *4*, thread suspension rollers.

measurement procedure involves an indication of the resonance frequencies f_0 and f_1 when both threads are placed in the cavity. The calculation algorithms for the characteristic impedance have the form [140]

$$\frac{r_{sh}}{Q} = \frac{\theta}{2\pi \mathcal{H}_f f_0} \frac{\Delta f_0}{f_0} \tag{6.17}$$

for a DLW and

$$\frac{r_{sh\,eff}}{Q} \approx \frac{\Delta f_0}{f_0} \frac{d_g}{\lambda_0} \left\{ \frac{\left[\sin\left[\dfrac{\pi d_g}{\lambda_0}\left(1+\dfrac{a}{d_g}\right)\right]\right]^2}{\dfrac{\pi d_g}{\lambda_0}} \right\} \left[\pi f_0 \mathcal{H}_f \left(1+\frac{2}{3}\frac{a}{d_g}\right)\right]^{-1} \tag{6.18}$$

for a BPS. The thread formfactor is determined in a reference cavity excited in the TM_{010} mode: $\mathcal{H}_f = 0.423\varepsilon_0 R_{ref}^2 \, \Delta f_{ref}/f_{ref}$.

In the range near $f_0 \approx 3$ GHz and for thread diameters of 0.2 and 1.0 mm, the deviation is equal to $\Delta f_0 \approx 1$ MHz and is independent of cavity length. This shift in resonance frequency can be measured on a general-purpose test bench with an accuracy of ±5% or better during several minutes.

A device to measure small shifts in the resonance frequency of a high-Q cavity can also be implemented using the phase lock principle. A block diagram of a measuring system of this type is shown in Fig. 6.5. When the resonance frequency of the cavity differs from the frequency of the measuring generator by $\Delta f_0'$, the phase shift

$$\Delta\varphi = \arctan\left(\frac{2Q_1}{f_0}\Delta f_0'\right)$$

occurs between couplers and the cavity. This shift is converted by a phase detector to an analog voltage signal. This amplified signal is used to change the frequency of the measuring generator, so as to maintain the state $\Delta\varphi \approx 0$ and $\Delta f_0' \approx 0$ in the system. The degree of the approximation $\Delta f_0' \approx 0$ depends on the characteristics of the system functional units: phase-detector resolution, analog-signal amplification, amplifier drift characteristics, an so forth.

For studying H-type cavities in the meter band, we compiled a measuring phase-lock setup entirely of high-quality mass produced instruments.

The general-purpose setup shown in Fig. 6.5 allows measurements of the natural frequency of a cavity, its deviation, and the loaded Q-factor. For $f_0 = 150$ MHz and $Q_1 = 6000$, this setup had the short-term instability of the measuring generator $\delta f_0/f_0 \approx 5 \cdot 10^{-8}$, which enabled us to measure the electric field parameter with an accuracy of $\Delta \xi_e/\xi_e \leq 10^{-2}$. The errors in determining the resonance frequencies and the loaded Q-factor were $\Delta f_0/f_0 \leq 3 \cdot 10^{-5}$ and $\Delta Q_1/Q_1 \leq 5 \cdot 10^{-2}$, respectively. The normal operation of the phase meter is maintained at a transition loss of the cavity $A \leq 30$ dB and, according to (6.6), at $Q_1 = 0.97 Q_0$.

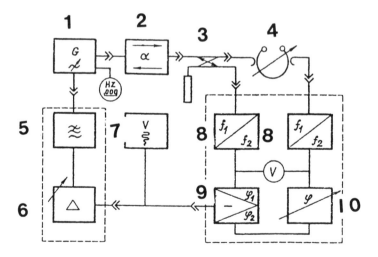

Figure 6.5. Block diagram of the general-purpose test system with a phase-locked loop: *1*, rf generator; *2*, attenuator; *3*, directional coupler; *4*, cavity; *5*, filter; *6*, amplifier; *7*, recording instrument; *8*, frequency divider; *9*, phase discriminator; and *10*, phase shifter.

The systematic error in phase-locked tracking the resonance frequency of the cavity is of the same character as in WBCTs and is determined by relation (6.16).

6.3 Measuring Probes

In relation (6.8), the resonance frequency change due to inserting a disturbing body into the cavity may be written in the form [4]

$$\mathcal{H}_i^{(e)} = \frac{\varepsilon_0 V_0}{4} \left(N_i + \frac{1}{\varepsilon - 1} \right)^{-1}, \qquad (6.19a)$$

$$\mathcal{H}_i^{(m)} = -\frac{\mu_0 V_0}{4} \left(N_i + \frac{1}{\mu - 1} \right)^{-1}, \qquad (6.19b)$$

where N_i are the diagonal elements of the electric or magnetic susceptibility tensor.

For probes made of perfect conductors, the formfactors have the form

$$\mathcal{H}_i^{(e)} = \frac{\varepsilon_0 V_0}{4} \frac{1}{N_i}, \qquad (6.20a)$$

$$\mathcal{H}_i^{(m)} = \frac{\mu_0 V_0}{4} \frac{1}{1 - N_i}. \qquad (6.20b)$$

Since the diagonal elements satisfy the condition

$$N_x + N_y + N_z = 1, \qquad (6.21)$$

then $N_x = N_y = N_z = 1/3$ for a spherical probe.

In this case, the corresponding formfactors are

$$\mathcal{H}_{x,y,z}^{(e)} = \pi \varepsilon_0 R_p^3, \quad \mathcal{H}_{x,y,z}^{(m)} = \frac{1}{2} \pi \mu_0 R_p^3,$$

where R_p is the radius of the spherical probe.

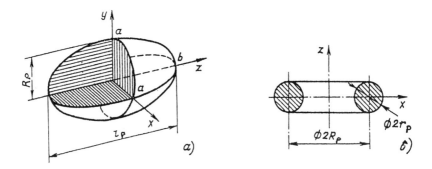

Figure 6.6. (a) Spheroid and (b) ring probes.

The formfactors of spheroids depend on their orientation in the field. Figure 6.6 shows a prolate spheroid and a coordinate system tied to it. For a prolate spheroid $(2R_p < l_p)$, we have [129]

$$N_z = \frac{1-\gamma^2}{\gamma^3}(\operatorname{arc\,tanh}\gamma - \gamma), \quad N_x = N_y,\qquad(6.22)$$

where $\gamma = [1 - (2R_p/l_p)^2]^{1/2}$. For an oblate spheroid $(2R_p > l_p)$,

$$N_z = \frac{1+\beta^2}{\beta^3}(\beta - \arctan\beta), N_x = N_y,\qquad(6.23)$$

where $\beta = [(2R_p/l_p)^2 - 1]^{1/2}$.

The probe directivity coefficient $\mathcal{H}_d^{(e,m)}$ is defined as the ratio of the maximum formfactor to its minimum value. For prolate and oblate spheroids, we have

$$\mathcal{H}_d^{(e)} = (1 - N_z)/2N_z,\qquad(6.24a)$$

$$\mathcal{H}_d^{(m)} = (1 + N_z)/[2(1 - N_z)].\qquad(6.24b)$$

Using the limit $(\operatorname{arctanh}\gamma - \gamma) \to [\ln(l_p/R_p) - 1]$ as $\gamma \to 0$, one can show that

$$\mathcal{H}_d^{(e)} = \left[8(R_p/l_p)^2 \ln(R_p/l_p) - 1 \right]^{-1}$$

for $2R_p/l_p \le 0.2$.

It follows that the electrical directivity coefficient of prolate spheroids can reach very large values. The magnetic directivity coefficient of oblate spheroids possesses a similar property. The directivity of spheroids makes it possible to determine the orientation of E and H vectors, which is necessary in studies of field topography of oscillations with an unknown wave type.

In practice, spheroid probes are not used because they are extremely difficult to manufacture. Instead, prolate $(l_p > 2R_p)$ and oblate $(2R_p > l_p)$ cylinders, called needles and discs, respectively are used. The formfactor of a needle probe can be estimated from Fig. 6.7, where more than 50 measured values $\mathcal{H}_z^{(e)} = F(2R_p/l_p)$ are plotted in the interval $0.05 \le 2R_p/l_p \le 0.7$. In the same figure, for metallic spheroids, theoretical curves $\mathcal{H}_d^{(e)}$ and $\mathcal{H}_z^{(e)}$ are worked out by formulas (6.20). (6.22), and (6.24).

A comparison between the calculated and measured data of Fig. 6.7 shows that the components of the electric susceptibility tensor of a prolate cylinder and those of a spheroid inscribed into

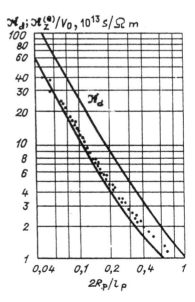

Figure 6.7. Theoretical curves for the formfactor and directivity coefficient of prolate conducting spheroids. Experimental values for cylindrical probes are shown by dots.

this cylinder are approximately equal. The ratio $N_z^{(\text{exp})}/N_z^{(\text{theor})} \approx$ 1.1 for $2R_p/l_p \approx 0.3$, and $N_z^{(\text{exp})}/N_z^{(\text{theor})} \to 1$ as $2R_p/l_p \to 0$. Similar calculated and measured data are shown in Fig. 6.8 for metallic discs and oblate spheroids with ratio $2R_p/l_p \geq 5$.

Thus, the formfactors of disc and needle probes can be estimated to sufficient accuracy by relations (6.20) with the values of N_i taken for the corresponding inscribed spheroids. A similar approach must be invoked to estimate the directivity coefficient.

For design of experiments and in approximate analysis of measured data, the following relations can be recommended:

$$\mathcal{H}_z^{(e)} \approx \frac{\pi \varepsilon_0}{4\sqrt{2}} l_p^2 \sqrt{R_p l_p}, \quad \mathcal{H}_d^{(e)} \approx 0.8 \left(\frac{2R_p}{l_p} \right)^{-3/2} \tag{6.25}$$

for metal needle probes if $2R_p/l_p \leq 0.4$ and

$$\mathcal{H}_z^{(m)} \approx \mu_0 R_p^3 \left[1 + \frac{4}{\pi} (2R_p/l_p)^{-1} \right], \quad \mathcal{H}_d^{(m)} \approx \frac{4}{\pi} \frac{R_p}{l_p} \tag{6.26}$$

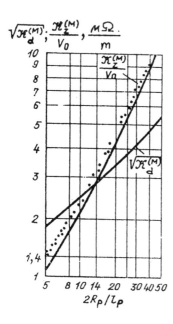

Figure 6.8. Experimental data for disc probes in a magnetic field. The solid lines show theoretical curves for inscribed spheroids.

for metal discs if $2R_p/l_p > 5$.

A ring probe shown in Fig. 6.6b is frequently used to measure magnetic and axisymmetric electric fields. A characteristic feature of ring probe theory is associated with the inductive interaction between ring probes and magnetic fields. The cross section of a ring is threaded by a variable flux of magnetic induction $d\Phi/dt = \mu_0 S H_z \omega_0$, where S is the ring area. The electromotive force $\mathcal{E} = -d\Phi/dt$ induces the current $I = \mathcal{E}/r_r = -(1/r_r)d\Phi/dt$. The magnetic energy stored in the ring is $\Delta W = LI^2/2 = (L/2)(1/r_r \ d\Phi/dt)^2$, where L is the ring inductance. The ring impedance is mainly inductive $(r_r = \omega_0 L)$. Consequently, $\Delta W = \mu_0^2 S^2 H_z^2 (2L)^{-1}$. Writing the theorem of small perturbations

$$\frac{\Delta f_0}{f_0} = \frac{1}{2}\frac{\Delta W}{W} = \mathcal{H}_z^{(m)}\frac{H_z^2}{W}$$

we find that

$$\mathcal{H}_z^{(m)} = \Delta W (2H_z^2)^{-1} = \mu_0 S^2/4L \ .$$

The ring inductance

$$L = \mu_0 R_p \left[\ln(8R_p/r_p) - 2\right] \approx \mu_0 R_p \ln(R_p/r_p),$$

and

$$S = \pi R_p^2 \ .$$

Finally, we have

$$\mathcal{H}_z^{(m)} \approx \frac{\mu_0 \pi^2 R_p^3}{4\ln(R_p/r_p)} \ . \tag{6.27}$$

The directivity coefficient of a ring in a magnetic field is

$$\mathcal{H}_d^{(m)} = \frac{\mathcal{H}_z^{(m)}}{\mathcal{H}_{xy}^{(m)}} = \frac{1}{3}\left(\frac{R_p}{r_p}\right)^2 \frac{1}{\ln(R_p/r_p)} \ . \tag{6.28}$$

A ring probe with $R_p/r_p > 5$ may be considered to be an infinite cylinder whose axis is perpendicular to the direction of electric field. In this case, $N_x = N_y \rightarrow 0.5$ and the formfactor in electric field is determined from (6.20):

$$\mathcal{H}_z^{(e)} = \varepsilon_0 \pi^2 r_p^2 R_p . \tag{6.29}$$

As follows from Fig. 6.9, the calculated characteristics of a ring probe (6.27) and (6.28) agree well with experimental data. The ring probe possesses a high directivity in a magnetic field ($\mathcal{H}_d^{(m)} = 65$ at $R_p/r_p = 25$) and a significant formfactor in an electric field ($\mathcal{H}_z^{(e)} = 1.8 \cdot 10^{-20}$ m²·s/Ω at $R_p = 5$ mm and $r_p = 0.2$ mm). These properties allow some measurements that cannot be achieved with spheroid probes.

When electric fields are examined near conducting surfaces, the probe formfactor increases due to the mirror-image effect. A rigorous quantitative analysis of this phenomenon meets with significant difficulties. For this reason, it is appropriate to use approximate methods and test their validity experimentally.

Let a spherical probe be placed near a conducting wall that serves as the boundary of a uniform field E_0. The effect of the wall on the electric field near the probe is taken into consideration by introducing a mirror image as shown in Fig. 6.10. The distribution of an electric field beyond a dielectric sphere placed in the uniform field has the form

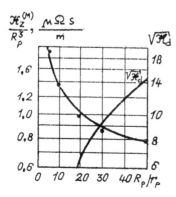

Figure 6.9. Ring-probe characteristics in a magnetic field. Experimental values are shown by dots.

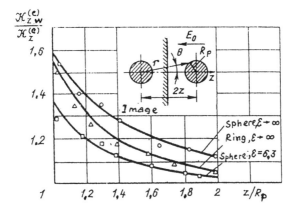

Figure 6.10. Conducting wall effects on the formfactors of ring and spherical probes. Experimental values are shown by dots.

$$E_r = -E_0 \left(1 + \frac{2R_p^3}{r^3} \frac{\varepsilon - 1}{\varepsilon + 2} \right) \cos\theta,$$

$$E_\theta = E_0 \left(1 - \frac{R_p^3}{r^3} \frac{\varepsilon - 1}{\varepsilon + 2} \right) \sin\theta,$$

$$(6.30)$$

where r and θ are the spherical coordinates whose origin is at the center of the probe's image. The probe will be in the image field E_i. This field squared can be found from (6.30) in the form

$$E_i^2 = E_r^2 + E_\theta^2$$
$$= E_0 \left[1 + \frac{2R_p^3}{r_p^3} \frac{\varepsilon - 1}{\varepsilon + 2} (2\cos^2\theta - \sin^2\theta) + \frac{R_p^6}{r^6} \left(\frac{\varepsilon - 1}{\varepsilon + 2} \right)^2 (4\cos^2\theta + \sin^2\theta) \right].$$

For further analysis, it is necessary to perform integration over the volume of the probe. Then

$$\int_{V_0} E_i^2 dV = 2\pi E_0 \int_0^{\arcsin(a/2z)} \sin\theta d\theta \int_{2z(\cos\theta - \sqrt{\cos^2\theta - k^2})}^{2z(\cos\theta + \sqrt{\cos^2\theta - k^2})} r^2 dr$$

$$\times \left[1 + \frac{2R_p^3}{r^3}\frac{\varepsilon-1}{\varepsilon+2}(3\cos^2\theta - 1) + \frac{R_p^6}{r^6}\left(\frac{\varepsilon-1}{\varepsilon+2}\right)^2 (3\cos^2\theta + 1)\right],$$

where $k^2 = 1 - [R_p/(2z)]^2$. Finally, we obtain

$$\int_{V_0} E_i^2 dV = V_0 E_0^2 \left\{ 1 + \frac{1}{2}\frac{\varepsilon-1}{\varepsilon+2}\frac{R_p^3}{z^3} + \frac{3}{2}\left(\frac{\varepsilon-1}{\varepsilon+2}\right)^2 \frac{R_p^3}{z^3}\right.$$

$$\times\left[\frac{1}{32}\left(\frac{\sinh 6M}{6} + \frac{\sinh 4M}{4} - \frac{\sinh 2M}{2} - M\right)\right.$$

$$\left.\left. + \frac{1}{24k^2}\left(\frac{\sinh 4M}{4} - \frac{\sinh 2M}{2}\right)\right]\right\}, \qquad (6.31)$$

where $M = \mathrm{arccosh}(1/k) = \ln[(1 + \sqrt{1 - k^2})/k]$. When the probe is located far from the wall $(R_p/r_p \to 0)$, the cavity frequency deviation $\Delta f_0/f_0 = \mathcal{H}^{(e)} E_0^2/W$. In the vicinity of the wall, we have

$$\frac{\Delta f_0^{(w)}}{f_0} \approx \frac{\mathcal{H}^{(e)}}{W}\frac{1}{V_0}\int_{V_0} E_i^2 dV,$$

where the approximate equality sign owes its existence to the replacement of the exact squared field by its average over the probe volume. A comparison between two latter formulas yields

$$\frac{\Delta f_0^{(w)}}{f_0} \approx \frac{1}{V_0 E_0^2}\int_{V_0} E_i^2 dV.$$

An increase in frequency deviation for the cavity with a uniform field may be interpreted as an increase in the probe formfactor, and the formfactor of a sphere near a conducting wall is

$$\mathcal{H}_w^{(e)} \approx \frac{\mathcal{H}^{(e)}}{V_0 E_0^2} \int_{V_0} E_i^2 dV . \tag{6.32}$$

Using the value of the integral

$$\int_{V_0} E_i^2 dV .$$

calculated above, we will consider two important cases.

(1) A spherical probe touches the wall ($z = R_p$);

$$\frac{\mathcal{H}_w^{(e)}}{\mathcal{H}^{(e)}} \approx 1 + \frac{1}{2}\frac{\varepsilon-1}{\varepsilon+2} + 0.137\left(\frac{\varepsilon-1}{\varepsilon+2}\right)^2 . \tag{6.33}$$

(2) A spherical probe is placed at distance $z \geq 2R_p$ from the wall;

$$\frac{\mathcal{H}_w^{(e)}}{\mathcal{H}^{(e)}} \approx 1 + \frac{1}{2}\frac{\varepsilon-1}{\varepsilon+2}\frac{R_p^3}{z^3} + \frac{1}{16}\left(\frac{\varepsilon-1}{\varepsilon+2}\right)^2\frac{R_p^6}{z^6}\left(1 + \frac{17}{60}\frac{R_p^2}{z^2}\right). \tag{6.34}$$

For a ring probe with dimension ratio $R_p/r_p \geq 5$, the dependence $\mathcal{H}_{zw}^{(e)}/\mathcal{H}_z^{(e)} = F(z)$ is simpler:

$$\frac{\mathcal{H}_w^{(e)}}{\mathcal{H}^{(e)}} \approx 1 + \frac{1}{2}\frac{\varepsilon-1}{\varepsilon+2}\frac{r_p^2}{z^2} + \left(\frac{\varepsilon-1}{\varepsilon+2}\right)^2\frac{r_p^4}{z^4}\left(4 - \frac{r_p^2}{z^2}\right)^{-2} . \tag{6.35}$$

Relations (6.33)–(6.35) lead to a number of implications important for metrology. A significant increase of the formfactor of a probe must be taken into account when fields are measured near metallic surfaces. If the probe touches the wall (without a conductive contact between them), then $\mathcal{H}_w^{(e)} \approx 1.64\mathcal{H}^{(e)}$ for a conducting sphere ($\varepsilon \to \infty$) and $\mathcal{H}_{zw}^{(e)} \approx 1.61\mathcal{H}_z^{(e)}$ for a ring probe.

The mirror-image effect is negligibly small ($\mathcal{H}_{zw}^{(e)}/\mathcal{H}_z^{(e)} < 1.01$) if the distance between the probe and the wall

$$z > 4R_p \text{ for a sphere and } z > 7r_p \text{ for a ring.}$$

Figure 6.10 shows the behavior of the ratio $\mathcal{H}_{z\,\mathrm{w}}^{(e)}/\mathcal{H}_z^{(e)}$ calculated for a spherical probe and a ring probe in the range $1 \le z/R_\mathrm{p} \le 2$. The same figure shows experimental data obtained with the WBCT-4 in a cavity excited in the TM_{010} mode. The approximate mirror-image theory is consistent with experiment not only qualitatively but also quantitatively. Therefore, its relations (6.33)–(6.35) may be used for practical estimations.

These relations determine probe formfactors accurate to within $\delta \mathcal{H}_{z\,\mathrm{w}}^{(e)}/\mathcal{H}_z^{(e)} = 10$–$20\%$. This error is mainly due to difficulties in measuring dimensions of small objects and in considering the defects of their shapes, their fastening on a thread, the presence of burrs, etc., rather than to inaccuracy of the theory. The dielectric permittivity of nonconducting probes is also known only roughly in most cases.

If probes that are manufactured according to calculated dimensions are used for absolute measurements of the parameters $\xi_\mathrm{e} = E/(PQ)^{1/2}$ and $\xi_\mathrm{m} = H/(PQ)^{1/2}$, their formfactors must be refined experimentally using a reference cavity. Methods of probe calibration and algorithms for probe formfactor calculation from the measured frequency deviation in the reference cavity excited in the TM_{010} mode are covered in our handbook [3].

A correct estimation of the measuring accuracy of the bead-pulling technique requires that not only instrumental errors but also specific errors due to field nonuniformity in the probe volume be taken into consideration. It is obvious that recommendations on the probe dimensions that ensure a specific accuracy of electromagnetic measurements present an important problem.

We will consider the solution to this problem for the practically important case of needle-probe measurements of the electric field strength parameter on the cavity axis where it is distributed by the law $\xi_\mathrm{e}(z) = \xi_0 \sin(z\,\theta/D)$.

The upper and lower bounds of the deviation of the resonant frequency of the cavity are defined as

$$\frac{\Delta f_0^{(max)}}{f_0} > \frac{\Delta f_0}{f_0} \ge \frac{\delta f_0}{2f_0}\left(\frac{\delta \xi_{\mathrm{e,m}}}{\xi_{\mathrm{e,m}}}\right)^{-1},\qquad (6.36)$$

where $\Delta f_0^{(max)}/f_0 \approx \pm 5 \cdot 10^{-4}$ is the frequency band of sustaining self-oscillations by a measuring system [see (6.13)], $\delta f_0/f_0$ is the short-term instability of self-oscillation frequency (6.11), and $\delta \xi_{e,m}/\xi_{e,m}$ is an accessible accuracy in field strength parameter measurements. Using the relation

$$\mathcal{H}^{(e,m)} = \frac{\Delta f_0}{f_0}\left(2\pi f_0 \xi_{e,m}^2\right)^{-1},$$

we can switch to the lower bound of the probe formfactor

$$\mathcal{H}^{(e,m)} \geq \frac{\delta f_0}{f_0}\left(4\pi f_0 \xi_{e,m}^2 \frac{\delta \xi_{e,m}}{\xi_{e,m}}\right)^{-1}. \tag{6.37}$$

The cavity must be replaced by a cavity-analogue excited in the same mode or the handbook values of $\xi_{e,m}$ should be used to estimate the formfactor quantitatively.

When the measurements of an accelerating electric field with TM_{010} oscillations are carried out at $r = 0$, relation (6.37) may be represented in the form

$$\mathcal{H}_z^{(e)} \geq \frac{\varepsilon_0 L}{16\pi}\left[\lambda_0 \nu_{01} J_1(\nu_{01})\right]^2 \frac{\delta f_0}{f_0}\left(\frac{\delta \xi_e}{\xi_e}\right)^{-1}$$

$$= 3.1 \cdot 10^{-2} \varepsilon_0 L \lambda_0^2 \frac{\delta f_0}{f_0}\left(\frac{\delta \xi_e}{\xi_e}\right)^{-1} \tag{6.38}$$

for a DLW mockup and in the form

$$\mathcal{H}_z^{(e)} \geq \frac{\delta f_0}{f_0} \frac{N_a}{L_a} \frac{\left[\lambda_0 \nu_{01} J_1(\nu_{01})(d_g + 2a/\nu_{01})\right]^2}{16\pi c Z_0}$$

$$= 2.74 \cdot 10^{-13} \frac{\delta f_0}{f_0} \frac{N_a}{L_a}\left[\lambda_0(d_g + 2a/\nu_{01})\right]^2 \tag{6.39}$$

for a BPS mockup. In the latter relation, N_a is the number of accelerating cells in the mockup, d_g is the length of an accelerating gap, and $L_a = D - 2t - L_c$.

In the case of H-type cavities with a cylindrical body,

$$\mathcal{H}_z^{(e)} \geq \frac{\delta f_0}{f_0} \frac{L \lambda_0 \varepsilon_0 (1 + 2.91 R^2 / L^2)}{60 \pi (k_g R \tilde{E}_z)^2}, \tag{6.40}$$

where k_g and E_z are found from the plots given in Chapter 5.

Field nonuniformity over the probe volume is described by the relation

$$\frac{\xi_e^{(\mathrm{meas})}(z)}{\xi_e(z)} \approx \left[\frac{1}{V_0 E_z^2(0,z)} \int_{V_0} E_z^2(r,z) dv \right]^{1/2}, \tag{6.41}$$

where $\xi_e(z)$ is the actual value of the electric strength parameter at coordinate z and $\xi_e^{(\mathrm{meas})}(z)$ is the measured value.

For a needle probe, the dependence on r in (6.41) is of no significance. For example, for cavities with TM_{010} oscillations, we have $\xi_{e0}(r_0) \sim E_0 J_0(2\pi r / \lambda_0)$ at the center of an accelerating gap, and

$$\frac{\xi_{e0}^{(\mathrm{meas})}(z)}{\xi_{e0}(z)} \approx \left[\frac{1}{E_0^2 \pi R_p^2 l_p} \int_0^{r_p} E_0^2 J_0^2 \left(\frac{2\pi}{\lambda_0} r \right) 2\pi l_p r dr \right]^{1/2}$$

$$\approx \left[1 - \left(\frac{\pi}{2} \right)^2 \left(\frac{R_p}{\lambda_0} \right)^2 \right]^{1/2}. \tag{6.42}$$

Relation (6.42) is equivalent to

$$\Delta \xi_{e0}^{(\mathrm{meas})} / \xi_{e0} = -\left(R_p / \lambda_0 \right)^2 \pi^2 / 8.$$

If we assume that $\Delta \xi_{e0}^{(\mathrm{meas})} / \xi_{e0} \leq 10^{-3}$, then $R_p / \lambda_0 \leq 2.8 \cdot 10^{-2}$, which is always satisfied for a needle probe. Therefore, only the dependence on z should be considered in a subsequent analysis of (6.41):

$$\frac{\xi_e^{(\text{meas})}(z)}{\xi_e(z)} \approx \left[\frac{1}{\pi R_p^2 l_p E_0^2 \sin^2(z\,\theta/D)} \int_{z-l_p/2}^{z+l_p/2} E_0^2 \sin^2(z\,\theta/D)\pi R_p^2 dz\right]^{1/2}$$

$$\approx \frac{\left[1 - \dfrac{\sin\varphi_p}{\varphi_p}\cos(2z\,\theta/D)\right]^{1/2}}{\sqrt{2}\sin(z\,\theta/D)}, \tag{6.43}$$

where $\varphi_p = \theta l_p/D$ is the angular size of the probe. In a field node $(z = 0)$,

$$\xi_e^{(\text{meas})}(0) = \frac{\xi_{e0}}{\sqrt{2}}\left(1 - \frac{\sin\varphi_p}{\varphi_p}\right)^{1/2} \approx \frac{\xi_{e0}\varphi_p}{\sqrt{12}}, \tag{6.44}$$

and in a maximum $(z = \pi D/(2\theta))$,

$$\xi_e^{(\text{meas})} = \frac{\xi_{e0}}{\sqrt{2}}\left(1 + \frac{\sin\varphi_p}{\varphi_p}\right)^{1/2} \approx \left(1 - \frac{\varphi_p^2}{24}\right)\xi_{e0},$$

which is equivalent to a systematic error in determining the field maximum

$$\Delta\xi_{e0}^{(\text{meas})}/\xi_{e0} = -\varphi_p^2/24. \tag{6.45}$$

From relations (6.44) and (6.45) it follows that the experimental pattern $\xi_{e0}^{(\text{meas})} = F(z_n)$ has the region of uncertainty at field zeros and involves a systematic underestimating of field amplitudes. These features are shown in Fig. 6.11, where an H-type cavity resonant in the TE_{11} mode is considered as an example.

A systematic difference between the measured and actual values of the field strength parameter, described by equation

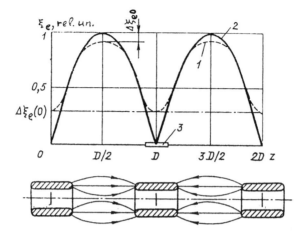

Figure 6.11. Effect of the needle-probe length on measuring the electric field on the axis of a cavity with drift tubes: *1;* experimental field pattern, *2,* field pattern as $\varphi_p \to 0$; and *3,* probe.

(6.43), leads to errors in determining cavity shunt impedances at $\theta_p = \sin\varphi_p/\varphi_p$:

$$\frac{r_{sh}^{(meas)}}{r_{sh}} = \left(\frac{U^{(meas)}}{U}\right)^2 = \left[\frac{\int_0^{\pi D/\theta_p}\sqrt{1-\theta_p\cos(2z\,\theta/D)}\,dz}{\int_0^{\pi D/\theta_p}\sin(2z\,\theta/D)\,dz}\right]^2, \qquad (6.46)$$

$$\frac{r_{sh\ eff}^{(meas)}}{r_{sh\ eff}} = \left\{\frac{2}{\pi}\left[\sqrt{\frac{1-\theta_p}{2}}+\frac{1+\theta_p}{2\sqrt{\theta_p}}\arcsin\sqrt{\frac{2\theta_p}{1+\theta_p}}\right]\right\}^2. \qquad (6.47)$$

The following criteria were proposed to choose the probe length:

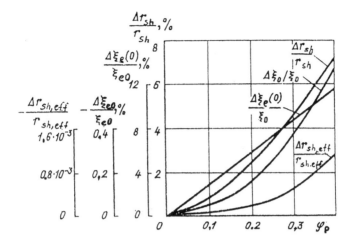

Figure 6.12. Nomograph for optimal probe dimensions.

(1) The region of uncertainty of field zeros must not exceed the random error in zero evaluation due to the frequency instability of instruments used. This requirement is satisfied if $\Delta\xi_e(0)/\xi_{e0} \leq 5\cdot10^{-2}$.

(2) The deficiency of the field amplitude must be negligibly small: $\Delta\xi_{e0}/\xi_{e0} \leq 5\cdot10^{-3}$.

(3) The systematic error in determining the shunt impedance from a measured field pattern must not exceed the rms error in measuring the Q-factor: $\Delta r_{sh}/r_{sh} \leq 2\cdot10^{-2}$.

An analysis of relations (6.44)–(6.47) plotted in Fig. 6.12 shows that all these criteria are fulfilled when $\varphi_p = \theta l_p/D < 0.2$ or $l_p \leq 0.2D/\theta$.

It is reasonable to use the term "optimal probe" for a needle probe having an angular size $\varphi_p \leq 0.2$ and satisfying the lower bound for the formfactor (6.37).

The reality of the optimal probe will be demonstrated with a typical example of the EDCs of the RELUS-1 accelerating subsection [141]. The initial data for this case are as follows: $\lambda_0 = 0.107$ m, $D = 5.36\cdot10^{-2}$ m, $L_a = 4.16\cdot10^{-2}$ m, $d_p = 2.36\cdot10^{-2}$ m, $N_a = 7$, $a = 5\cdot10^{-3}$ m, $\delta\xi_{e0}/\xi_{e0} \leq 10^{-2}$, and $\delta f_0/f_0 \leq 5\cdot10^{-7}$. Substituting these data into (6.39), we obtain that $\mathcal{H}_z^{(e)} \geq 2.05 \cdot 10^{-20}$ m^{2}s/Ω, so that, in view of (6.25), we find $l_p \geq 2.13 \cdot 10^{-3}$ m. The angular size of the probe $\varphi_p = \pi \cdot 2 \cdot 13/53.6 = 0.13$; therefore, this is an optimal probe.

The concepts proposed for calculations of a probe's optimum size ensure a high accuracy ($\Delta \xi_{e0}/\xi_{e0} \leq 10^{-2}$ and $\Delta r_{\text{sh}\cdot\text{eff}}/r_{\text{sh}\cdot\text{eff}} \leq 2 \cdot 10^{-2}$) without invoking a special processing of experimental data set and present one of the most important points of the procedure under consideration.

6.4 Measurement Procedure and Data Processing

Measurements of axial electric fields in accelerating structures are most important of all and require highest accuracy, nevertheless they are easy to perform. Recommendations on how to conduct such measurements with WBCT setups are given in our papers [3, 122]. The most tedious and complicated experiments relate to the topography of electromagnetic fields in accelerating structures and to identification of modes within a wide frequency band.

It is expedient to conduct these experiments in two stages. In the first stage, the resonance frequencies of mockups containing $N = 2$, 3, and 4 structure cells are successively determined in a given frequency band. Two half-cells at the ends are treated as one cell. In the search for the resonance frequencies, it should be taken into account that the $\theta = 0$ mode of the transverse electric field is not excited in such mockups. For a preliminary assessment of weather or not close resonances belong to one and the same mode, the data obtained must be plotted in the coordinates $f-\theta$ disregarding the sign of dispersion. The mode is first determined from the relation $\theta = \pi n / N$. If $N = 2$, then $\theta = 0$, $\pi/2$, and π. If $N = 3$, then $\theta = 0$, $\pi/3$, $2\pi/3$, and π. If $N = 4$, then $\theta = 0$, $\pi/4$, $\pi/2$, $3\pi/4$, and π. Frequencies that coincide in all three cases relate to the $\theta = 0$ or π mode. Identical frequencies for the first and third cases relate to the $\theta = 0$, $\pi/2$, or π mode. This selection rule enables one to reveal the $\theta = 0$, $\pi/2$, and π modes and also to plot other resonances in the f–θ diagram (without considering the sign of dispersion). In the case when either the frequency bands of different modes overlap or mixed dispersion occurs, the measurements of resonance frequencies of a six-cell mockup may be required. In this case, the selection rule is supplemented by modes that are multiple of $\pi/3$.

For determining the sign of dispersion and the topography of electromagnetic field, it is recommended to perform the measurements on a mockup equipped with a set of probes as shown in Fig. 6.13. A needle probe *1* mounted on a rotating fluoroplastic disc *2* determines a mode type (TM or TE) and the number of field variations in the azimuth. A ring probe *3* mounted on the same disc is used to indicate the component of magnetic

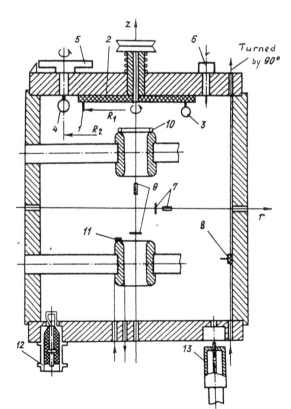

Figure 6.13. Jigs to study the topography of the electromagnetic field in the resonant CRS mockup.

field $H_\varphi(R_1; \varphi; 0)$. The direction of the vector $\boldsymbol{H}(R_1; \varphi; 0)$ is found with a ring probe 4 mounted on a rotating holder 5. The holder is inserted into short-circuit plane holes drilled at $\Delta\varphi = 15°$ intervals with radius R_2. Unused holes are stopped with plug 6. The radial distribution of the components E_z or E_r is measured by a needle or disc probe 7 moving along a thin nylon thread. If the polarization of field is unknown, the probes 7 are doubly moved at mutually perpendicular positions of threads. The distribution $E_r(R, \varphi_p, z)$, where φ_p is the angular coordinate of the polarization plane, is measured by a needle probe fixed on a foam-plastic holder 8. The holder fixes a needle orientation along a cavity radius. If the design of an accelerating structure does not allow probe movement along the z axis at $r = R$ (DLW, DAW), it is also necessary to carry out the measurements of $E_z(0, 0, z)$ and $E_r(0, 0, z)$ with a needle and a disc 9, respectively.

Additionally, operation-mode measurements of the electric field strength parameter in the vertices of the drift tubes must be performed with a ring (10) or plate (11) probe.

The replacement of the loop couplers 12 by the stub antennas 13 aligned with the axes of the cavity significantly facilitates the selection of close resonances related to different modes. In this case, TM_{0np} modes are primarily excited in the cavity.

An analysis of the results of these measurements identifies the type of modes (TM, TE, or TEM wave) and phase advance θ, and synthesizes the topography of electromagnetic field.

By way of example, Fig. 6.14 presents the results obtained for a resonance mockup of a conducting rod structure (CRS). The topography of electromagnetic field is consistent with all measured field patterns and corresponds to a transverse wave with $\theta = \pi/2$ in a coaxial line.

The investigation of CRSs by the procedure proposed revealed 17 frequency bands in the frequency range 1–5 GHz, identified modes and the sign of dispersion within each frequency band, and determined the coupling impedance at the synchronous points ($\beta_{ph,n} = 1$) of eight higher order modes (HOMs). The data obtained were used to design a variant of accelerating section of the storage accelerator [123].

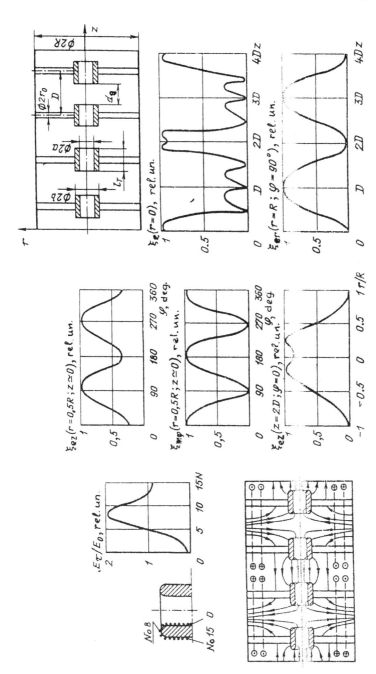

Figure 6.14. Synthesis of the topography of CRS electromagnetic field from experimental data. The cavity dimensions are $2R = 129$ mm, $D = 69$ mm, $l_{dt} = 26$ mm, $2b = 29.5$ mm, $2a = 22$ mm, $2r_0 = 15$ mm; $\varphi = 0$ in the stem plane; and $f_{\pi/2} = 1091$ MHz.

Measured data on resonance frequencies, Q-factor, and cavity transition attenuation obtained on a semi-automatic test bench are processed using conventional techniques. For this purpose, three to five frequency readings at a corresponding response amplitude on the display or three to five marks of the position of a calibrated attenuator are taken. The measured data are averaged, and a maximum random error or (for a large number of measurements) the rms error is calculated. Frequency measurements are reduced to the room temperature $T = 20°C$.

Fields measured with WBCT-type setups are printed out as an log containing the numbers of experimental points and the corresponding resonance frequencies of a cavity. The first step of processing lies in calculating the frequency deviation $\Delta f_n = |f_0 - f_0(n)|$. The unperturbed frequency f_0 is taken from the mean frequency reading in the case when the probe is located beyond the cavity. If an optimal probe is used, the mean value of frequencies corresponding to probe positions at field nodal points may be taken as f_0. Furthermore, the electric ($\xi_e(n)$) or magnetic ($\xi_m(n)$) field parameter is calculated by relation (6.9) at each experimental point. The effective shunt impedance or the amplitude of the fundamental harmonic of accelerating field is determined by the numerical integration of the pattern $\xi_e = F(n)$ by the method of trapezoids:

$$
\frac{r_{\text{sh eff}}}{Q} = \left[\sum_1^{N_a} \frac{2}{\lambda_w} \int_{-\lambda_w/4}^{\lambda_w/4} \xi_e(z) \cos \frac{2\pi z}{\lambda_w} dz \right]^2
$$

$$
\cong \left[\frac{\Delta z_n}{L} \sum_1^{N_a} \xi_e \frac{z_n + z_{n+1}}{2} \cos \left(\frac{\pi}{D} \frac{z_n + z_{n+1}}{2} \right) \right]^2, \qquad (6.48)
$$

$$
\xi_{e0} = \frac{4}{\lambda_w} \int_{-\lambda_w/4}^{\lambda_w/4} \xi_e(z) \cos \frac{2\pi z}{\lambda_w} dz
$$

$$
\cong \frac{4\Delta z_n}{\lambda_w} \sum_1^{N_a} \xi_e \frac{z_n + z_{n+1}}{2} \cos \left[\frac{\pi}{\lambda_w} (z_n + z_{n+1}) \right], \qquad (6.49)
$$

where N_a is the number of accelerating cells in a BPS mockup.

Numerical examples of processing experimental data of the form $\xi_e = F(_n)$ by relations (6.48) and (6.49) are given in [4].

If the parameter ξ_{e0} of the investigated cavity does not allow the use of an optimal-sized probe, the measured values $\xi_e(n)$ must be corrected taking into account the field nonuniformity over the probe length. A correction may be required in studies of extended cavities or in the cases when the field pattern contains very steep segments. The latter case can occur in BPSs and H-type cavities, where the function $\xi_e(z)$ is close to an exponential function within drift tubes. The following procedure is developed to process this type of measured data.

In the neighborhood of a point with number n, the field pattern is represented by the first three terms of the Taylor series:

$$\xi_e(z) = \xi_e(z_n) + (z - z_n)\xi_e'(z_n) + \frac{(z - z_n)^2}{2}\xi_e''(z_n). \tag{6.50}$$

For correcting $\xi_e^{(meas)}(n)$, it is necessary to calculate the integral

$$\int_{z_n - l_p/2}^{z_n + l_p/2} \xi_e^2(z)dz = l_p\left\{1 + \frac{l_p^2}{12}\left[\frac{\xi_e''(n)}{\xi_e(n)} + \left(\frac{\xi_e'(n)}{\xi_e(n)}\right)^2\right] + \frac{l_p^4}{320}\left(\frac{\xi_e''(n)}{\xi_e(n)}\right)^2\right\}. \tag{6.51}$$

Using the experimental points adjacent to point n, we find

$$\xi_e'(n) = \frac{\xi_e^{(meas)}(n+1) - \xi_e^{(meas)}(n-1)}{2\Delta z_n},$$

$$\xi_e''(n) = \frac{1}{\Delta z_n}\left[\frac{\xi_e^{(meas)}(n+1) - \xi_e^{(meas)}(n)}{\Delta z_n} - \frac{\xi_e^{(meas)}(n) - \xi_e^{(meas)}(n-1)}{\Delta z_n}\right].$$

$$\tag{6.52}$$

Substituting (6.52) into (6.51) and (6.51) into (6.50), we obtain the correction formula

$$\xi_e(n) = \xi_e^{(meas)}(n)\left[1 + \frac{1}{12}\left(\frac{l_p}{\Delta z_n}\right)^2 (\delta_1 + \delta_2)\right]^{-1/2}. \qquad (6.53)$$

where

$$\delta_1 = \frac{1}{4}\left(\frac{\Delta\xi_e^{(meas)}(n+1,n-1)}{\xi_e^{(meas)}}\right)^2$$

and

$$\delta_2 = \frac{\Delta\xi_e^{(meas)}(n+1,n) - \Delta\xi_e^{(meas)}(n,n-1)}{\xi_e^{(meas)}(n)}.$$

By way of example, we consider the correction of measurements of the field on the axis of an accelerating BPS cell that was used to test calculated reference data [4]. This cell is characterized by $f_0 = 3$ GHz, $D = 50$ mm, and $\Delta z_n = 1.56$ mm. The measurements were carried out with an optimal probe ($l_p = \Delta z_n$). The results of these measurements on the pattern interval $-\Delta z_n \leq z \leq D/2$ are summarized in Table 6.1.

The last column of Table 6.1 presents a typical situation occurring in experimental studies of cavities with drift tubes. The fronts of the pattern $\xi_e(n)$ are systematically overestimated, its bend regions are underestimated, and its region near the gap is measured without distortions. The correction of $\xi_e^{(meas)}(n)$ with the optimal probe is no greater than 5%, and the systematic errors in the effective shunt impedance is no greater than 0.5% and the systematic error $\Delta r_{sh}/r_{sh} \approx 1.5$–2%. It may be inferred that the correction of measurements with the optimal probe is only required in exceptional cases. One of these cases relates to calculations of particle dynamics in compact proton accelerators with sign-variable rf focusing, where the tolerable error $\delta\xi_e/\xi_e$ is 1–2% [142].

To estimate the systematic error $\xi_e^{(meas)}(n)$ on the fronts of the electric-field pattern in cavities with drift tubes, it is convenient to use the relation

Table 6.1. Processing of measured data for the electric field strength parameter

n	z_n, mm	$\xi_e^{(\text{meas})}$, $\Omega^{1/2}/\text{m}$	δ_1	δ_2	$\xi_e(n)$ $\Omega^{1/2}/\text{m}$	$\dfrac{\delta\xi_e^{(\text{meas})}}{\xi_e^{(\text{meas})}}$, %
0	24.96	0	0	0	0	0
1	23.40	14	1.06	0.07	13.4	4.3
2	21.84	29	0.60	0.52	27.8	4.1
3	20.28	59	0.56	0.47	56.6	4.1
4	18.72	117	0.47	0.38	113.0	3.4
5	17.16	219	0.32	0.20	214.4	2.1
6	15.60	364	0.14	−0.047	362.6	0.4
7	14.04	492	0.048	−0.083	492.7	−0.1
8	12.48	579	0.015	−0.055	580.0	−0.2
9	10.92	634	0.003	−0.066	635.7	−0.3
10	9.36	647	0	−0.029	647.8	−0.1
11	7.80	641	0	0	641.0	0
12	6.24	635	0	−0.016	635.4	−0.06
13	4.68	619	0	0.015	618.6	−0.06
14	3.12	612	0	0.002	612.0	0
15	1.56	606	0	0.007	606.0	0
16	0	604	0	0	604.0	0
17	−1.56	606	0	0	606.0	0

$$\frac{\xi_e}{\xi_e^{(\text{meas})}} = \left[\frac{\sinh\left(v_{01}l_p/a\right)}{v_{01}l_p/a}\right]^{-1/2}. \tag{6.54}$$

Applying (6.54) to the data of Table 6.1 at $v_{01}l_p/a = 0.75$, we obtain $\xi_e = 0.995\,\xi_e^{(\text{meas})}$ and $\delta\xi_e^{(\text{meas})}/\xi_e^{(\text{meas})} = 4.5\%$.

Relation (6.54) serves as a criterion for the choice of a procedure used to process on-axis electric-field data measured in accelerating structures with drift tubes. If the systematic error calculated from this relation is greater than the acceptable error, the processing must be performed using the correcting relation (6.53).

6.5 Tuning of Accelerating Cavities

Accelerating structures whose dimensions have been determined by calculations or using reference data [4] may have manufactured cells differing in resonance frequencies. These differences change the electric field distribution over the structure length. The may be due to calculation errors or an imperfect manufacturing technique. For example, with ±0.02 mm rms tolerances on linear dimensions, the relative difference in the resonance frequencies of DLW and BPS cells at a frequency of 3 GHz will be ±10⁻³ and ±3·10⁻³, respectively. Even with a perfect brazing technique, an accelerating structure assembled of such cells will need subsequent tuning.

Tune-up procedures for DLWs, including for those with variable cell dimensions, are thoroughly described in [3, 4] and have come into common use at present. Tune-up procedures for BPSs are less known.

If the frequency of the π/2 mode is specified as the operating frequency, it follows from (4.45) that the dispersion characteristic of a BPS with accelerating cells at the ends is continuous provided that the resonance frequencies of the cells with perturbations due to coupling slots taken into account satisfy the conditions

$$f_1 = f_N = f_{\pi/2}\sqrt{1 - k_c(2p-1, 2p+1)/2},$$

$$f_{2p+1} = f_{\pi/2}\sqrt{1 - k_c(2p-1, 2p+1)}, p = 1,2,3,...,(N-3)/2 \tag{6.55}$$

for the accelerating cells and the conditions

$$f_2 = f_{N-1} = f_{\pi/2}\sqrt{1 - k_c(2p, 2p+2)/2},$$

$$f_{2p} = f_{\pi/2}\sqrt{1 - k_c(2p, 2p+2)}, p = 2,3,4,...,(N-3)/2 \tag{6.56}$$

for the coupling cells.

Continuous cell numbering is used in relations (6.55) and (6.56). Odd numbers correspond to accelerating cells, and $k_c(2p - 1, 2p + 1)$ and $k_c(2p, 2p + 2)$ are the cross-coupling coefficients of

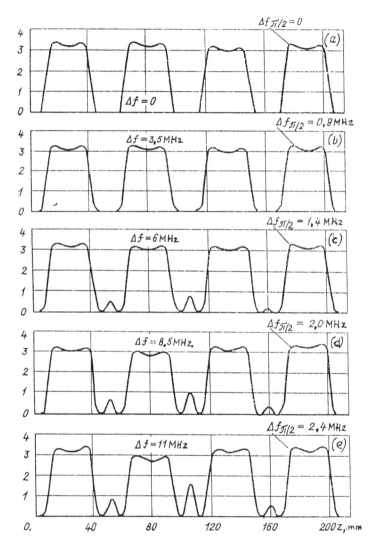

Figure 6.15. Evolution of the BPS accelerating field pattern due to detuning of the resonance frequency of an accelerating cavity ($N_a = 4$).

single-type cells. The cross-coupling coefficients of an on-axis coupled biperiodic structure or a BPS with coaxial coupling cells are negligibly small if the coupling slots stand at 180° relative to each other. It follows from (6.55) and (6.56) that all the cells must be tuned to the frequency of the $\pi/2$ mode. Deviations of cell resonant frequencies from a nominal value lead to the onset of electric field on the axis of the coupling cells and to the redistribution of electric fields over the accelerating cells. Quantitatively, this phenomenon is shown in Fig. 6.15. The relationship between the electric field strengths at the centers of adjacent accelerating cells depends also on their coefficient of coupling with a common coupling cell:

$$\frac{E_{2p-1}}{E_{2p+1}} = \frac{k_c(2p, 2p+1)}{k_c(2p-1, 2p)}. \tag{6.57}$$

The variances of relative nonuniformity of accelerating field σ_1^2 and σ_2^2 and the variances of relative deviations of the coupling coefficients $\sigma^2(\Delta k_c/k_c)$ and of the resonance frequencies of cells $\sigma^2(\Delta f_a/f_{\pi/2})$ and $\sigma^2(\Delta f_c/f_{\pi/2})$ are related as follows [143]:

$$\left.\begin{aligned}
\sigma_1^2 &= \frac{N_a+1}{6}\sigma^2\left(\frac{\Delta k_c}{k_c}\right), \\
\sigma_2^2 &= \frac{4(N_a+1)N_a}{3}\left(\frac{2}{k_c}\right)^4 \sigma^2\left(\frac{\Delta f_a}{f_{\pi/2}}\right)\sigma^2\left(\frac{\Delta f_c}{f_{\pi/2}}\right),
\end{aligned}\right\} \tag{6.58}$$

where N_a is the number of accelerating cells in the structure.

An increase in the effective shunt impedance of a BPS due to differing resonance frequencies of accelerating cells is determined by the relation

$$\frac{\Delta r_{\text{sh eff}}}{r_{\text{sh eff}}} = \frac{8N_a}{k_c^2}\frac{Q_a}{Q_c}\sigma^2\left(\frac{\Delta f_a}{f_{\pi/2}}\right). \tag{6.59}$$

Dependence (6.59) is plotted in Fig. 6.16 for $\Delta r_{\text{sh·eff}}/r_{\text{sh·eff}} \leq 0.05$. The ordinate is the maximum allowable frequency deviation $\Delta f_a/f_{\pi/2} = 3\sigma$ for a normal deviation distribution. An increase in the

Figure 6.16. Maximum deviation of the resonance frequencies of accelerating cells for $\Delta r_{\text{sh·eff}}/r_{\text{sh·eff}} \leq 5 \cdot 10^{-2}$.

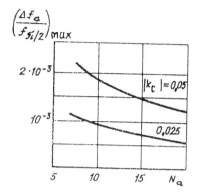

effective shunt impedance within 5% is reached when $(\Delta f_a/f_{\pi/2})_{\max} \leq 0.7 \cdot 10^{-3}$ or $\Delta f_{\pi/2} = \pm 2 \text{ MHz}$ at $f_{\pi/2} \approx 3 \text{ GHz}$. The maximum allowable deviation of the coefficient of coupling between adjacent cells (see (6.58)) at $(\Delta E/E)_{\max} = \pm 5$ and $\pm 2\%$ can be estimated from Fig. 6.17. Calculations of electron dynamics have shown that, for sections with $\beta_{\text{ph}} = 1$, the spread $(\Delta E/E)_{\max} \approx 5\%$ has little effect on the output characteristics of a beam. Therefore, the following tolerance on BPS tuning by the coefficient of coupling between adjacent cells may be used: $|\Delta k_c (2p - 1.2p)/k_c| \leq 0.05$. If $(\Delta f_a/f_{\pi/2})_{\max} \leq 0.7 \cdot 10^{-3}$, it follows from the second equation of system (6.58) that the maximum allowable spread in the resonance frequencies of coupling cells is at least three times greater; i.e., $(\Delta f_c/f_{\pi/2})_{\max} \leq 2.1 \cdot 10^{-3}$. Thus, the tolerances

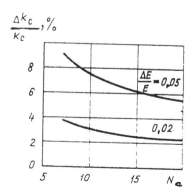

Figure 6.17. Maximum deviation of the coefficient of coupling between BPS cells.

on tuning the resonance frequencies of the cells of a one-tank SW electron linac are not too close.

The tolerances on the difference of the resonance frequencies of accelerating sections Δf_c and on the mean coefficient of coupling between the accelerating sections $\Delta k_c^{(s)}$ of two-section electron linacs with a bridge feeding network [97] are determined from the condition of compensated reflections at the operating frequency mode $\pi/2$ and its closest modes. For a stable one-frequency excitation of a magnetron, it is necessary to ensure the fulfillment of the conditions

$$\frac{\Delta f_c}{\Delta f_{\pi/2}} \leq 0.7 \cdot 10^{-4} \text{ and } \left| \frac{\Delta k_c^{(s)}}{k_c} \right| \leq \frac{\Delta f_c}{f_{\pi/2}} \frac{f_{\pi/2}}{F_0} |,$$

where F_0 is the frequency pulling coefficient of the magnetron. If $F_0 = 10$ MHz and $f_{\pi/2} \approx 3$ GHz, then $|\Delta k_c^{(s)}/k_c| \leq 2.1\%$. Careful tuning of accelerating sections makes it possible to fulfill these very stringent requirements for their identity.

A $\pi/2$ mode on-axis coupled structure is tuned on a semi-automatic test bench using post couplers. The tuning diagram is shown in Fig. 6.18, where couplers 1 and 7 are placed in a position to measure the resonance frequency of an accelerating cell 4. According to equation (3.6), the resonance frequency of a pth cell perturbed by coupling slots is measured at $H_{p-1} = H_{p+1} = 0$, i.e., when adjacent cells are not excited. The adjacent cells are short-circuited by the bars carrying the stub antennas, which significantly increases their resonant frequency. As a result, the condition $H_{p-1} = H_{p+1} = 0$ proves to be fulfilled automatically. Before tuning, the coordinate z_{st} is chosen such that the transition attenuation through the cells be about 30 dB as recommended in Chapter 3. The stub antennas are moved relative to end plates 2 and 6 and are placed successively in drift tubes 1–2, 2–3, 3–4, etc. The resonance frequencies of the accelerating cells 3 and 4 and coupling cells, as well as their Q-factors, are measured in each position. When the stub antennas are in drift tubes 1–3 and 2–4, the 0 and π modes are measured, which enables one to calculate the coefficient of coupling between adjacent cells as $k_c(p, p + 1) =$

$(f_\pi^2 - f_0^2)/(f_\pi^2 + f_0^2)$. The resonance frequencies of the cells are corrected by changing their dimensions. If the resonance frequencies are greater than their rated values, the diameter of the cells is increased. If the resonance frequencies are smaller than their nominal values, the coupling cells are tuned by means of facets 5 shown in Fig. 6.18 and the accelerating cells are tuned by reducing the length of the drift tubes l_{dt}. The coupling coefficient of cell pairs is tuned to the greatest measured value by means of a small increase in the aperture angle of coupling slots. In the course of tuning, the derivatives of frequency with respect to dimensions $\partial f_0/\partial q_i$ must be used.

The resonance frequency of the cell with power input (coupler) is measured when a matched load is connected to a rectangular waveguide. The set of these procedures enables tuning the resonance frequencies of the cells with deviations $\Delta f_c/f_{\pi/2} = \Delta f_a/f_{\pi/2} \leq$ $2 \cdot 10^{-4}$. For accelerating sections of RELUS linacs, these deviations are no greater than 10^{-4}.

BPS assembly with the use of vacuum-tight brazing is accompanied by the deformation of the walls between accelerating and coupling cells due to residual-stress relief during annealing, which results in detuned resonance frequencies of the cells. The

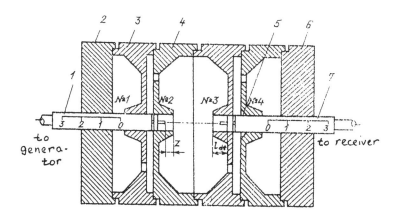

Figure 6.18. BPS equipment for tuning the resonance frequencies of cells.

violation of technical guidelines may also lead to solder wicking into the cells. A BPS tuning procedure that follows brazing involves measurements of the resonance frequencies, as illustrated in Fig. 6.18, and their correction by means of the device shown in Fig. 6.19. This device is inserted into a tuned cell through a transit channel of the structure, and its grip *1* is fixed by spacer *2*. A wall deflection is made by weak blows upon the bar carrying the grip. Figure 6.19 shows the device in a position when the deflection of wall *3* is possible, which corresponds to an increase in the resonance frequency of the coupling cell and to a decrease in that of the accelerating cell. The deflection required is chosen in a few attempts, each followed by a resonance-frequency measurement. The order of tuning is determined from an analysis of the table listing the resonance frequencies of the cells after brazing. However, in any case, one should start the tuning from the cells with the greatest frequency deviations. RELUS-1 and RELUS-2 accelerating sections were tuned by this procedure with a spread in the resonance frequencies from 0.7 to −0.5 MHz, while the operating-mode frequencies of the sections were identical to an accuracy of ±0.1 MHz.

Bunchers are tuned to a given distribution of electric field by matching the coupling coefficients of their adjacent cells through the change of the aperture angle ψ of coupling cells. It follows from (6.57) and (3.10) that

$$\frac{E_{2p-1}}{E_{2p+1}} = \left(\frac{\psi_{2p,2p+1}}{\psi_{2p-1,2p}} \right)^3 . \tag{6.60}$$

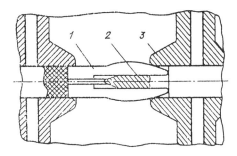

Figure 6.19. Device for correcting the resonance frequencies of cells in a brazed BPS.

Figure 6.20. RELUS-1 accelerator buncher.

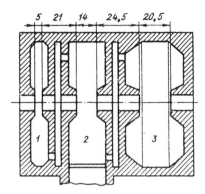

The tuneup is illustrated in Figs. 6.20 and 6.21, which show the diagram of the RELUS-1 buncher and the process of tuning to a given ratio of the electric fields at the centers of accelerating cells *1* and *3* E_3/E_1 = 5.3. In the first stage, the resonance frequencies of buncher cells are tuned to a nominal value with accuracy $| \Delta f_{a,c} | \leq$

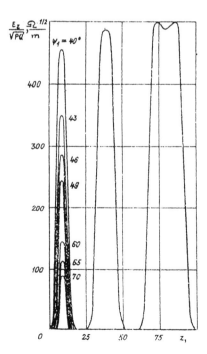

Figure 6.21. RELUS-1 buncher tuning to a specified distribution of the electric field.

(0.3–0.4)·$10^{-3}f_{\pi/2}$ and the pattern of axial electric field is measured. This stage relates to $\psi = 40°$ in Fig. 6.21. The angle of the coupling slot between the first accelerating cell and the adjacent coupling cell is then estimated by formula (6.60), which converts to

$$\psi_{2p-1,2p} = \psi_{2p,2p+1}\sqrt[3]{\frac{E_3 E_1^{(1)}}{E_1 E_3^{(1)}}}\,,\qquad(6.61)$$

where $E_3^{(1)}/E_1^{(1)}$ is the initial ratio of the electric fields in cells 3 and 1. Substituting the value $E_3^{(1)}/E_1^{(1)} = 1.077$ obtained from the pattern of Fig. 6.21, $\psi_{2p,2p + 1} = 40°$, and $E_3/E_1 = 5.3$, we find that $\psi_{1,2} = 68°$. The aperture angle of the coupling slot between the first accelerating cell and the adjacent coupling cell is increased successively approaching its calculated value. Each boring of the slot is followed by tuning the resonance frequencies of the cells to the frequency $(f_{\pi/2} \pm 1)$ MHz and by measuring the pattern of electric field. The tuning process is shown in Fig. 6.21. The desired ratio $E_3/E_1 = 5.3$ is reached at $\psi_{1,2} = 70°$, which is close to the calculated value. A large number of steps in changing the dimensions of coupling cells are made here to test the procedure.

BPSs designs denying a reduction of the second nearest neighbor couplings to a negligible level must be tuned by a procedure developed for annular-coupled structures and used in the RELUS-2 accelerator [144].

The tuneup starts from coupling cells 1 in the stack of cells (mockup) shown in Fig. 6.22. This mockup includes modules 2 and 3 of the structure and the end accelerating half-cells 4 and 5 connected to the modules. Bars 6 and 7 of a coupler are mated in the connector plane of modules 2 and 3. The resonance frequency of a coupling cell is then measured, and the cell is tuned to a given value with accuracy $\Delta f_c/f_{\pi/2} < 10^{-3}$. When the resonance frequencies of all coupling cells are tuned, the resonance mockup shown in Fig. 6.23 is assembled. This mockup contains two pairs of modules 4–5 and 6–7 that form two preliminary tuned coupling cells 1 and 3, an accelerating cell 2, and two end accelerating cells 8 and 9. Bars 10 and 11 of couplers are inserted into a transit channel up to the connector planes of cells 1 and 3 so that exciting

Figure 6.22. Tuning of an annular coupling structure with the second nearest neighbor coupling taken into account.

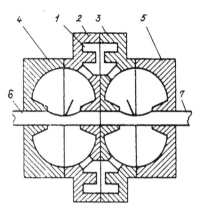

(*12*) and receiving (*13*) antennas (see Fig. 6.25) be directed to coupling slots. The pairs of modules *4–5* and *6–7* are turned relative to one another in the connector plane *A–A* until the π/2-frequency of the mockup is equal to the resonance frequency of the preliminary tuned coupling cells *1* and *3*. Subsequent measurements and structure assembly for brazing are performed at the chosen relative positions of the modules.

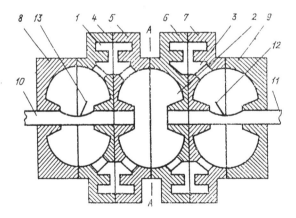

Figure 6.23. Tuning of an accelerating cell of an annular coupled structure with the second nearest neighbor coupling taken into account.

The frequency of the accelerating cell 2 is defined through the 0-, $\pi/2$-, and π-frequencies of the mockup from the set of equations

$$\left.\begin{array}{l} \Delta_0 = 0, \\ \Delta_{\pi/2} = 0, \\ \Delta_\pi = 0, \end{array}\right\}, \qquad (6.62)$$

where Δ_0, $\Delta_{\pi/2}$, and Δ_π are the determinants of the set of equations for three coupled resonance circuits. For example, in the case of the 0-mode, the determinant has the form

$$\Delta_0 = \begin{vmatrix} 1 - \dfrac{f_1^2}{f_0^2} & \dfrac{k_{12}}{2} & \dfrac{k_{13}}{2} \\ \dfrac{k_{12}}{2} & 1 - \dfrac{f_2^2}{f_0^2} & \dfrac{k_{23}}{2} \\ \dfrac{k_{13}}{2} & \dfrac{k_{23}}{2} & 1 - \dfrac{f_3^2}{f_0^2} \end{vmatrix}, \qquad (6.63)$$

where f_1 and f_3 are the resonance frequencies of the coupling cells 1 and 3, respectively; f_{02} is the desired resonance frequency of an accelerating cell; and k_{12}, k_{13}, and k_{23} are the coefficients of coupling between cells 1–2, 1–3, and 2–3, respectively, and $k_{13} = 0$ due to the turn of the modules considered above.

The frequencies of the end accelerating cells are determined in the resonance mockup shown in Fig. 6.24. Here, coupler 4 is made as in Fig. 6.18 and the 0 and π modes of the mockup are measured. The resonance frequency of the end accelerating cell 1 is calculated from the set of equations

$$\left.\begin{array}{l} \Delta_0 = 0, \\ \Delta_\pi = 0. \end{array}\right\} \qquad (6.64)$$

The determinants Δ_0 and Δ_π have the form

Figure 6.24. Measurements of the resonance frequencies of the end accelerating cells.

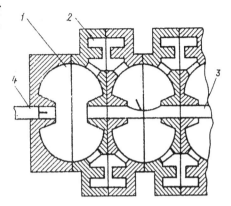

$$\Delta_0 = \begin{vmatrix} 1 - \dfrac{f_1^2}{f_0^2} & \dfrac{k_{12}}{2} \\[2ex] \dfrac{k_{12}}{2} & 1 - \dfrac{f_2^2}{f_0^2} \end{vmatrix}, \qquad (6.65)$$

where k_{12} is the coefficient of coupling between cells *1* and *2*.

Brazing is made once the coupling cells *2* have been tuned, the mutual orientation of each pair of the modules has been determined, and the accelerating cells have been tuned. After brazing, the resonance frequencies of the cells are measured and tuned with the device shown in Fig. 6.19 in the same order as before brazing.

Figure 6.25 shows coupler *3* used to realize the tune-up procedure under consideration. A stub antenna *1* is brazed to the inner conductor *2* of a coaxial cable *3*. When installed in the transit channel, the antenna is placed in groove *5* made in a tubular bar *4*. The stub antenna is made of elastic bronze wire and has a spring compensator *6* providing the device's durability. Plug *7* excludes a direct coupling between antennas in the tuning of coupling cells.

It follows from (6.50) for structures with large coupling coefficients $(k_c \approx 40\%)$ that tolerances on the resonance frequencies and coupling coefficients of cells become much less

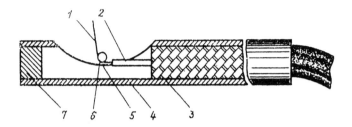

Figure 6.25. Design of the coupler for tuning an annular coupled structure.

stringent and can be achieved at the stage of design of accelerating sections. A biperiodic accelerating structure of this type, for example, a DAW, does not require individual tuning of cells.

7

LINEAR
RESONANT
CAVITY
ACCELERATORS

7.1 Background

In preceding chapters, we covered methods that refer to calculations and measurements of accelerating cavities in RELUS standing-wave electron linacs with BPSs [97, 145] and the Uragan-2 proton linac with accelerating inter-digital cavities excited in the TE_{111} mode [146]. This chapter describes these accelerators and the design and tuneup of their cavities.

The advantages of the standing-wave mode over the traveling-wave mode in the design of small-size low-energy electron linacs are well known [14, 147, 148]. Using relations (4.3) and (4.47), one can construct the dependence shown in Fig. 7.1, which displays the normalized energy as a function of the attenuation parameter $\tau = \alpha L$ for different linacs. The ordinate is the maximum increment of the kinetic energy $\Delta W_k = eU$ (e is the electron charge) normalized by $(r_{sh.eff}PL)^{1/2}$ or $(r_{sh}PL)^{1/2}$.

237

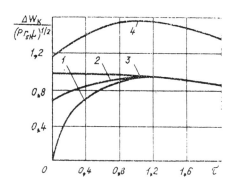

Figure 7.1 Normalized growth of energy as a function of the attenuation parameter: *1*, traveling-wave electron linac; *2*, standing-wave electron linac ($\pi/2$ mode), *3*, standing-wave electron linac ($\theta = \pi$); and *4*, linac with beam recirculation at two circulations.

It is seen that the standing-wave regime is preferable in high-gradient compact machines with high gradients of the accelerating field when low rf powers are used. In such small-length systems, the rf power fed from the generator is multiply reflected and accumulated within the structure.

It is clear that a TW linac can be made as compact as a SW linac only by substantially increasing the electric field strength at the expense of a significant decrease in the iris aperture. As a result, the group velocity decreases, waveguide dimensions are subject to closer tolerances, and beam guiding problems arise.

A comparative analysis of 5–10 MeV TW and SW linacs with an average beam power of tens of kilowatts, and thus with a high intensity (Chapter 1), also demonstrates advantages of using BPSs for these purposes.

First Russian RELUS-type SW electron linacs with BPSs were designed in the Moscow Engineering Physics Institute (MEPI) for energies ranging from 3 to 10 MeV and pulse currents up to 100 mA. In their design, we implemented a program that involves the following stages:

— a feasibility study of standing-wave electron linacs in industry and medicine and the choice of an accelerating structure [145];

— theoretical studies and the development of a software package to calculate the dynamic range of BPSs, the longitudinal and radial dynamics of particles in SW linacs with beam loading and transient processes taken into account;

— the design of instrumentation for experimental studies of
  BPSs and the development of relevant tuning techniques;
— the choice couplers for SW linacs, their feasibility study and
  technical and experimental verification;
— the design, fabrication, startup, and performance studies of
  linacs with different of accelerating structures and couplers.

We consider here a 0.5-MeV proton linac designed in MEPI to
illustrate the design and tuneup of inter-digital H-type cavities
[142]. This machine has a low injection energy (30 keV) and a
rather low rate of energy accumulation. These features imply that
this system is radiation safe, and can be used, in particular, for
ion implantation in semiconductors and integrated circuits. The
accelerator has a small size due to a sign-variable rf focusing of
protons in drift tubes and a rather low rf pulse power (below
35 kW).

7.2 RF Feeding Circuits of RELUS Accelerators

Linac performance characteristics are largely determined by the
type of rf generators and feeding circuits of accelerating sections.
A magnetron-type generator suits best for multipurpose accelerators
because of its high efficiency, low cost, rather long life,
availability, and a relatively low supply voltage. A disadvantage
of magnetrons for SW linacs lies in the difficulty of attaining
stable operation for the rf section of the accelerator with a high Q-
factor if special care is not taken. The point is that such linac
circuits are characterized by a high frequency sensitivity of the
tank input impedance, the presence of reflected waves in the rf
sections under transient conditions (the total reflection of rf
power from the structure is observed at the first turn-on of the rf
source), spurious modes near the operation mode of the tanks,
and a strong effect of section beam loading on the rf source.

Some feeding circuits of SW linacs use either a waveguide
bridge or a ferrite isolator or circulator. In the first case, rf power
is fed to two subsections with identical resonance frequencies, Q-
factors, and coupling with transmission lines, through a 3-dB

hybrid junction. Since no reflections occur in the input branch of this junction, the magnetron will operate stably onto a matched load. However, no frequency stabilization is achieved in this case [149]. To attain the required stabilization, it is necessary to detune the resonance frequencies of the tanks (f_{01}, f_{02}) or to imbalance the junction.

The condition for the absence of spurious modes at frequencies close to the operation frequency can be found from

$$\frac{\partial}{\partial f}(B_m + B_1) > 0 , \tag{7.1}$$

where B_m and B_l are the reactive conductances of the magnetron and loading. From (7.1), we obtain [150]

$$\Gamma_{in,max} \leq 1 + \frac{Q_t}{Q_{ext,m}} - \sqrt{\left(1 + \frac{Q_t}{Q_{ext,m}}\right)^2 - 1} , \tag{7.2}$$

where $\Gamma_{in,max}$ is the maximum input reflection coefficient, $Q_t = 2\pi(L/\lambda)(\lambda_{wg}/\lambda)$ is the Q-factor of the transmission line, L is its length, and $Q_{ext,m}$ is the external Q-factor of the magnetron.

The input reflection coefficient depends on the subsection frequency detuning and on the outer power ratio of the junction as follows:

$$\Gamma_{in} \approx \frac{Q_{ext}\delta_0}{(1+G)^2 + (Q_{ext}\delta_0)^2} \tag{7.3}$$

if $Q_{ext}\delta_0 < 1$ and $Q_{ext}\delta_0 = 0$ and

$$\Gamma_{in} \approx 2\Delta k_b^2 \tag{7.4}$$

if $Q_{ext}\delta_0 < 1$ and $Q_{ext}\delta_0 \gg 1$. Here, $G = Q_{ext}/Q_0$ is the active conductivity of the subsection at the input of the transmission line; Q_{ext} and Q_0 are the external and unloaded Q-factors of the subsection; and $\delta_0 = (f_{01} - f_{02})/f_0$, where $f_0 = (f_{01} + f_{02})/2$ and k_b is the bridge coupling coefficient.

If we restrict ourselves to $\Gamma_{\text{in,max}} \leq 0.2$, then we can find the transmission line length from (7.3) and the tolerable resonant frequency detuning δ_0 and bridge imbalance Δk_b from (7.3) and (7.4):

$$Q_{\text{ext}}\delta_0 \leq 2.5 - \sqrt{6.25 - (1+G)^2} \,, \tag{7.5}$$

$$\Delta k_b^2 \leq 0.1. \tag{7.6}$$

At a bridge unbalance Δk_b, the magnetron stabilization coefficient k_{st} for identical subsections can be obtained by the formula

$$k_{\text{st}} = 1 - \frac{Q_{\text{ext}}}{Q_{\text{ext,m}}} \frac{8}{(1+G)^2} \frac{\Delta k_b}{k_b}. \tag{7.7}$$

An analysis of the dependence of the stabilization coefficient on matching conditions shows that the optimum conditions are better attained by frequency detuning within the limits controlled by (7.5) rather than by junction imbalance.

In RELUS accelerators, rf power is fed through waveguide bridges (RELUS-1, 2, 3) or through a ferrite isolator (RELUS-4). For the circuit with a ferrite isolator, the magnetron stabilization coefficient is written as [126]

$$k_{\text{st}} = 1 + \frac{8\Phi\beta_0 Q_l F_0}{[\beta_0(1-\Phi) + (1+\Phi)]^2 0.834 f_0}. \tag{7.8}$$

Here, F_0 is the magnetron frequency pulling band; β_0 is the coefficient of coupling between an accelerating cavity and a coupler; $\Phi = 10^{-(A_1+A_2)/20}$; A_1 and A_2 describe the power losses within the isolator in the forward and backward directions, respectively; and Q_l is the loaded Q-factor of the cavity.

The frequency-capture band in the stabilized self-oscillation mode is found from the condition

$$\frac{\Delta f_c}{f_0} = \frac{\beta_0}{4Q_{\text{ext,m}}} \left[1 - \left(\frac{1-\Phi}{1+\Phi}\right)^2\right]\left(\beta_0 \frac{1-\Phi}{1+\Phi} + 1\right)^{-1}. \tag{7.9}$$

For the magnetron-fed RELUS-4, $f_0 = 2.8$ GHz, $F_0 = 8$ MHz, $Q_{ext,m} = 150$, $Q_1 = 1.3 \cdot 10^4$, $\beta_0 = 2$, and $\Phi = 0.2$; we have $k_{st} = 19$ and $\Delta f_c = \pm 2.2$ MHz. Under these conditions, the magnetron can be easily brought into the regime of self-oscillations stabilized by an external cell by manual frequency tuning.

An analysis of magnetron operation on a high-Q resonance load such as a RELUS-4 accelerating cavity, but connected through a ferrite circulator, may be found in the paper of Kaminskii and Milovanov [151].

The advantages of a circuit feeding a SW electron linac through a ferrite isolator include its simplicity, compactness, and the possibility of its assembly from standard units. The main disadvantage of this circuit is associated with feeding power losses in a direct passage through the isolator. A circuit that feeds accelerating sections through a waveguide bridge has an efficiency of about 100%; however, it is more sophisticated and its design and manufacture are very expensive. The latter disadvantage is eliminated by mass production of such circuits.

Thus, the circuits feeding accelerating sections through a ferrite isolator and a waveguide bridge do not offer clear advantages over each other. For this reason, both feeding circuits were used and tested in designing RELUS-type standing-wave electron linacs.

Schematic diagrams of the rf sections of RELUS accelerators are shown in Fig. 7.2.

The rf section of the RELUS-1 with a self-contained buncher feeder (Fig. 7.2a) is the most complicated design. This section is made of an ordinary 72×34 mm rectangular waveguide, so that the magnetron power is fed through a non-reflecting coupler from a circular to a rectangular waveguide. The power is then transmitted through a phase shifter, the straight branches of a 7.8-dB directional coupler, a controllable attenuator, a vacuum window, into a bridge with a division factor of 2.6 dB. The accelerating sections are connected to the output bridge branches. The part of the rf power, which is separated from the main section by the 7.8-dB directional coupler, is fed into the buncher through the phase shifter and the vacuum window. The power reflected

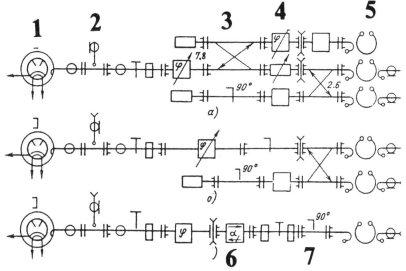

Figure 7.2 Schematic diagrams of the rf section of RELUS accelerators: (a) RELUS-1, (b) RELUS-2 and RELUS-3, and (c) RELUS-4. *1*, magnetron; *2*, waveguide coaxial coupler; *3*, directional coupler; *4*, phase shifter; *5*, accelerating cavities; *6*, ferrite isolator; *7*, waveguide rotation by 90°.

from the accelerating and bunching sections during a transient process is absorbed in loads connected to the corresponding branches of the bridge and directional coupler.

The tanks and a part of the rf section are evacuated by waveguide-vacuum collectors whose positions are shown in Fig. 2a. A rotation by 90° in the E-plane allows a higher circuit density for the section on the platform. The configuration of the RELUS-1 setup is typical for a multipurpose accelerator and offers a wide tuning of beam parameters by changing both the feeding power of the accelerating sections and the phase of bunched beam entrance into the accelerating sections. These tunings are independent, which is especially important for a comprehensive study of the performance of a test accelerator specimen and for a comparison of calculated and measured data with different initial conditions.

The design of an rf section for RELUS-2 and RELUS-3 setups (Fig. 7.2b) includes the same units except for an independent feeding circuit of the buncher, which is integrated with the first accelerating section. The accelerator offers the tuning of beam parameters at the output by changing the rf power and by detuning the resonance frequencies of accelerating sections relative to each other. A high circuit density of the section and a low mass of its units allows accelerator placement on a rotating platform.

The RELUS-4 rf section in Fig. 7.2c corresponds to a special-purpose electron linac and is very simple. A ferrite isolator is mounted in a 72 × 44 mm rectangular waveguide and is connected to the section through adapters. A phase shifter is made as an insert whose length is chosen during adjustment and remains unchanged. The design section units is discussed below.

7.3 Accelerating Cavities for RELUS Accelerators

The design of DLWs with specified laws of phase-velocity and electric-field variation along the axis are discussed in depth in our handbooks [3, 4].

The geometric dimensions are calculated so accurately that the allowance for the inner diameter $\Delta(2b)/2b \approx -2 \cdot 10^{-3}$ is sufficient for tuning the assembled system to the specified characteristics.

To illustrate the method of BPS design, which forms the basis for the design of RELUS accelerators, it is expedient to consider a few specific examples. We illustrate this method by using an electron linac for radiation technology as an example. The accelerator is characterized by an energy of 5 MeV at $I = 0.1$ A; its feeding source is a 1.5-MW pulsed magnetron operating at a frequency of 2797 MHz. It is desired to have minimal longitudinal size of the setup, low production cost, simple operation, and high reliability. The calculation procedure is summarized in Table 7.1.

Table 7.1 Flow chart of engineering design of a series of BPSs

Design stage	Decision or result
1. Choose a configuration meeting the criteria of the request for proposal	Single-tank accelerator driven via a ferrite valve
2. Choose a BPS satisfying the requirements of the request for proposal from reference data.	BPS with internal cylindrical coupling cells optimized in the effective shunt impedance
3. Recalculate geometrical dimensions and EDCs of an optimized accelerating cell for the operating wavelength with the scaling factor $M = \lambda_0/\lambda_{ref}$.	$\lambda_0 = 107.2$ mm. The profile of the first accelerating cell from a reference data bank. The recalculation coefficient $M = 1.0715$. One module of the structure is depicted in Fig. 7.3; $r_{sh,eff}(k_c = 0) = 77$ MΩ/m, $k_{ov} = 4.4$.
4. Estimate the length of the structure and the number of accelerating cells from the initial data with allowance for the rf power loss in the feeding line. Relations (4.47) and (4.49).	$P_0 = 1.3$ MW, $L = 0.5$ m, $W_k = 7$ MeV at $I = 0$; $W_k = 5.4$ MeV at $I = 0.1$; $N_a = 10$.
5. Check the electrical strength of the structure by the relation $E_{max} = k_{ov}(r_{sh,eff}P_0/)^{1/2}$	$E_{max} = 61.6$ MV/m
6. Evaluate the coupling coefficient from the condition of stable single-frequency excitation of the magnetron $$\|k_c\| \geq 2F_0\left[f_{1/2}\cos\frac{\pi(N_a-2)}{2(N_a-1)}\right]^{-1}$$	For the frequency pulling coefficient $F_0 \approx 10$ MHz, $\|k_c\| \geq 0.041$

Table 7.1 Continued

7. Correct $r_{sh,eff}$ for the coupling factor with the formula (4.71). Correct the energy and β_0 with formulas (4.47) and (4.49). Determine the final number of accelerating cells in a tank.	$r_{sh,eff}(k_c = -0.041) =$ 71 MΩ/m, $W_k = 5.2$ MeV at $I = 0.1$ A, $\beta_0 = 1.7$ and $N_a = 10$ remain as before.
8. Calculate the geometrical dimensions of coupling slots with formulas (3.10), (3.11), (3.13), (3.14) or (3.19) $$l_{sl} = \left[\frac{6Z_0|k_c|}{Z_{sl}H_c(r_{sl})H_a(r_{sl})} \right]^{1/3}$$ The length of a slot with rounded short sides $$l_{sl,r} = l_{sl}\left(1 + 0.155\sin\frac{\pi\Delta}{l_{sl}}\right)$$	Initial data [mm]: $\Delta = 4$, $t = 2.14$, $L_a = 45$, $L_c = 4.3$, $r_{sl} = 27.45$, $R = \lambda_{1/2}v_{01}/2\pi = 41$. Calculation: $H_a = 71.28$ m$^{-3/2}$, $H_c = 230.61$ m$^{-3/2}$, $l_{sl} = 28.6$ mm, $l_{sl,r} = 29.5$ mm, $\psi = 61°30'$ (see Fig. 7.3).
9. Determine the diameter of the coupling cell with allowance for the effect of coupling slots and beam holes $$2R_c = \frac{2R + 2(\Delta f_c + \Delta f_{c,h})}{\partial f_c/\partial R},$$ $$-\Delta f_c = f_{1/2}\left[1 - \left(1 - k_c\sqrt{L_a/L_c}\right)^{-1/2}\right],$$ $$\Delta f_{c,h} = 0.358N_0 f_{1/2}(a/R)^2 \, a/L_c,$$ where N_0 is the number of beam holes.	$\Delta f_c = -169$ MHz, $\Delta f_{c,h} = 45.9$ MHz. $\partial f_c/\partial R = -68$ MHz/mm, $2R_c = 78.3$ mm.
10. Refine the diameter of accelerating cells [by formula (3.10)] and the derivative $\partial f_a/\partial R$	$\Delta f_a = -17.6$ MHz, $\partial f_a/\partial R = -62$ MHz/mm, $2R_a = 79.66$ mm.

Table 7.1 Continued

11. Determining the width of the rf input matching aperture by formula (3.46)	Inputs: $A \times B = 72 \times 34$ mm, $Q = 14 \times 10^4$, $Z_{in} = 1.7$, $t = 1$ mm, $h = 20.9$ mm.
12. Determine the internal diameter of an rf input cell [by formula (3.51a)] with allowance for the difference of its profile from the accelerating cells and the effect of a coupling slot	$\Delta f_c = 17.7$ MHz, $\partial f_c/\partial R = -62$ MHz/mm, $\Delta 2R = -0.57$ mm, $2R_{cc} = 2R_a - 0.57$ mm $-$ 1.88 mm $= 77.2$ mm. See Fig. 7.4 for an input cell.
13. Recommend a manufacturing technology for designers of detail drawings.	Surface finish of live surfaces $R_a = 0.32$ μm. The roughness height of brazed surfaces $R_a = 1.25$ μm. The runout of the module connectors shall be within 0.02 mm. The flatness of the side walls to be within 0.01. All the angular dimensions to be within ±20'. All linear dimensions, except $2R_a$ and $2R_c$, to be within ±0.01 mm. The tolerance on R_r is ±0.05 mm.

Note. The cells (modules) of accelerating systems shall be made of oxide free copper in two steps. In the first step, all parts are manufactured with an allowance of 0.5–0.6 mm. In the second step, they are annealed to relieve stress and machined to within the given tolerances. Recognizing the approximate character of the correcting formulas, it is expedient to reduce the diameters $2R_a$ and $2R_c$ by 0.5–0.6 mm and machine accurate to ±0.05 mm. Brazing is performed by 72% silver solder in a hydrogen atmosphere or in a vacuum. The quality of brazing

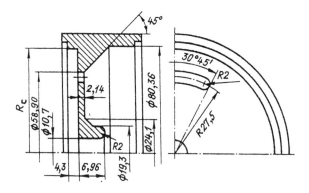

Figure 7.3 Accelerating cell and a coupling cell.

Creating RELUS-type accelerators in practice was preceded by theoretical studies in the dynamics of electrons in the field of a BPS standing wave. The most important components of the software package used were the Luch-24 and Luch-50 codes to calculate longitudinal motions of electrons in axisymmetric fields with current loading included [61] and the Radius code to calculate electron radial motions [152]. Calculations by these codes, along with studies of BPS EDCs, resulted in the formulation of the request for proposal of RELUS-type accelerators, which is summarized in Table 7.2.

Smaller trapping coefficients in Table 7.2 correspond to the operation mode in the absence of an external focusing field. Accelerating sections of RELUS setups are made on the basis of BPSs with an efficient shunting resistance of 61–70 MΩ/m and provide a high rate of energy gain – up to 9 MeV/m for RELUS-4.

Table 7.3 lists the section characteristics obtained from certifying measurements. The general diagram of bunchers is

---

is especially important for H-type cavities which pass considerable currents though joints of stems with drift tubes and the body. When the live surfaces are machined with the roughness height $(0.25–0.2)\delta$, where δ is the skin depth (1.10), and the modules are brazed by silver, the quality coefficient of BPSs and H-type resonators attains a value of 0.9

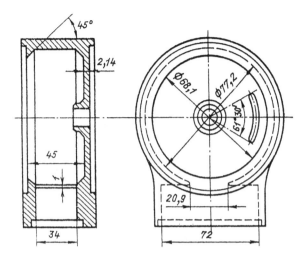

Figure 7.4 RF input cell.

shown in Fig. 7.5, and their calculated EDCs and dimensions are summarized in Table 7.4.

The design of accelerating sections was developed in the course of tests (including at a high power level) with a few mockups differing in assembly and in the configuration of units and parts.

The final accelerating structure design that may be recommended for full-scale production conditions is shown in Fig. 7.6. Functionally, the design of an acceleration structure includes the following components: an accelerating cavity, a power input unit, a cooling system, signal terminals, and connectors.

For two-section setups, this set must be complemented by a device to tune the resonance frequency of a section.

The accelerating cavity is built of modules of two standard dimensions (7 and 8 in Fig. 7.6) and is completed at the ends by cells 18. The water cooling system of pipe-inside-pipe type is formed by two housing halves 9 and 22, an inlet pipe union 2, an outlet union 11, and canals made in the flanges. Hot stream flow

Table 7.2 Characteristics of RELUS linacs

	RELUS-1	RELUS-2	RELUS-3	RELUS-4
Energy, MeV	4.5	5	3.2	10
Rated current, mA	60	50	100	20
Regulation range, MeV	2.5–5.1	3–6	1.8–3.4	5 and 10
Max. current, mA	115	120	200	120
Pulse beam power, kW	320	350	300	350
Energy spectrum, %	5–20	5–20	20	2–5
Output beam diameter, mm	4	4	6	4–6
Capture coefficient, %	18–29	20–40	30	35
Magnetron frequency stabilization factor, minimum	10	11	10	15
RF power, MW	1.4	1.4	1.4	1.4
RF pulse width/repetition frequency, μs/Hz	3/400	3/400	3/400	3/400
Max. external magnetic focusing field, A/m	7.2×10^4	9.6×10^4	no field	7.2×10^4
Injection voltage, kV	40	40	40	40
Bremsstrahlung at 1 m from target	420	400	250	650

from the left half of the system to its right half is provided by longitudinal canals drilled in the housing of the power feeding unit *4*. A signal terminal (view *A*) is connected to the housing of this unit, whose surface is not concealed by the water cooling

Table 7.3 Characteristics of accelerating cavities of RELUS linacs

	RELUS-1	RELUS-2	RELUS-3	RELUS-4
Number of tanks	3	2	2	1
Number of buncher cells	3	2	2	2
Coupling cell type	internal	external	internal	external prismatic
Length of structure, m	0.85	0.72	0.41	0.8
Effective shunt impedance, MΩ/m	62	67	61	69
Unloaded Q	13×10^3	14.6×10^3	13×10^3	14×10^3
Maximum electric field strength on the axis, kV/cm	200	220	240	220
Coupling coeff., %	2.4	5	3	6.5
Overvoltage factor	4.02	2.8	4	2.8
Diameter over the water cooling casing, mm	115	1160	115	120
Max. acceleration rate, MeV/m	13	32	115	32
Coefficient of coupling with the rf waveguide	2.1	2.2	1.6	2.2

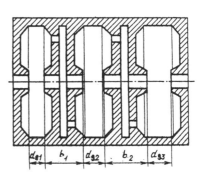

Figure 7.5. Sectional view of a buncher.

Table 7.4 Characteristics of bunchers of RELUS linacs

	d_{cap1}, mm	d_{cap2}, mm	d_{cap3}, mm	dg_1, mm	dg_{cap}, mm
RELUS-1	5	15.3	23.7	21	23.1
RELUS-2	21.4	27.5		14.8	20.1
RELUS-3	13.5	20.5			
RELUS-4	19.6	21.2		17.5	20.5

	β_{ph1}	β_{ph2}	β_{ph3}	$E_1{:}E_2{:}E_3$	E_3, kV/cm
RELUS-1	0.34	0.67	0.93	0.17:1:1	220
RELUS-2	0.665	0.864	1	0.16:1.19:1	116
RELUS-3	0.67	0.78	1	0.5:0.67:1	240
RELUS-4	0.655	0.75	1	0.86:1:1	187

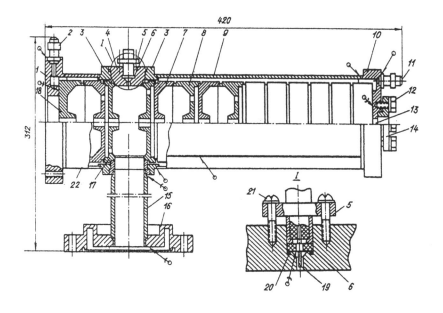

Figure 7.6 Accelerating section of a RELUS linac.

housings. A transition attenuation of the coupler of about 70 dB is provided by a stub antenna *19* placed in the beyond-cutoff canal with a diameter of about $0.04\lambda_{1/2}$. Coupler vacuum gasketing is provided by a ceramic insulator *6* and an indium gasket *20*. Gasket squeezing is implemented via the housing of rf connector *5* by screws *21*.

The connectors are made in the form of flanges: an input flange *1* with a neck for the metallic vacuum gasket of the mating flange of the injector, buncher, or first section; a throttle flange *16* to connect the section to the waveguide feeding circuit; a terminal flange *10*; and a thickening flange *14* that provides the vacuum tightness of the window through which an accelerated beam is extracted. This window is overlapped by partition *13* made of aluminum foil with thickness $t_f = 0.2$ mm. The threads of the terminal flange *10*, which are released when bolts *12* are taken apart, are used to connect the accelerator to the vacuum electron duct (e.g., to the chamber of magnetic energy analyzer).

The modules of the accelerating cavity *7* and *8*, the housing of the power feeding unit, the rectangular waveguide *15*, and coupler diaphragms *17* are made of oxygen-free copper (MOB). The other parts are made of chromium stainless steel. This choice of materials ensures both the compatibility of temperature deformations during soldering in a hydrogen atmosphere and high corrosion-resistant properties of the setup.

The housings of the water cooling system are built of two semitubes each and are welded on the shoulders of flanges *1* and *10* and plates *3* after vacuum control and the tuning of the resonance frequencies of the cells of the slow-wave system. The semitubes are joined to one another by a longitudinal seam shown in Fig. 7.6. The device for tuning resonance frequencies is also joined to the housing of the power feeding unit. The vacuum tightness of the device and tuning element motion are provided by a sylphon.

A correct calculation of the cooling system for accelerating sections is very important for their design. Insufficient cooling may lead to the unbalance of structure resonance frequencies and to a decrease in the effective shunting resistance. Excessive cooling is responsible for a large consumption of water and increases operational costs, especially in the absence of return water.

It should be emphasized that the concepts proposed to design accelerating structures based on BPSs are largely derived from the experience in designing traveling-wave electron linacs.

7.4 Design and Characteristics of RELUS Linacs

A rational linac design is an important factor in the practical realization of linac design advantages. RELUS-1, 2, and 3 linacs are stationary systems. RELUS-4 is a mobile unit. For the first three units, it was important to minimize the space occupied by the machine. In the last unit, the designer aimed at achieving a densely configured assembly with the least weight of components.

The space occupied by a linac is minimized by adopting a vertical-plane design. The arrangement of RELUS-1 is schematized in Fig. 7.7. A vertical plane design is achieved by an appropriate orienting the rf power inputs into accelerating sections and buncher and the use of a special 3dB bridge and directional coupler with coupling over the wide wall. Evacuating means are arranged in a second vertical plane so that their projection overlaps a side protruding part of the attenuator. RELUS-1 (without modulator) has a footing of $0.84 \, \text{m}^2$, RELUS-2 has a foundation of only $0.58 \, \text{m}^2$.

Physically, RELUS accelerators combine the following functional systems: acceleration, rf power supply, injection, rf feeding, vacuum, hydraulic, and measurement and control. The machine has a short length and low weight, because it is a SW design using BPS and high acceleration rate with a limited rf power. It takes advantage of using a waveguiding bridge and a directional coupler with coupling over a wide wall.

A long service life and a high reliability of this type of linac is achieved by using materials with high corrosion and radiation resistance (oxide-free copper, alloyed steel, rolled aluminum, ceramic materials, etc.), improved life of standard components,

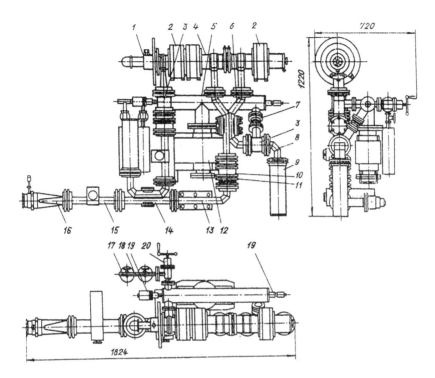

Figure 7.7. Arrangement of a RELUS accelerator: *1*, injector; *2*, focusing coils; *3*, waveguide manifold; *4*, accelerating system; *5*, vacuum manifold; *6*, waveguide bridge; *7*, vacuum bellows; *8*, waveguide bend; *9*, absorbing load; *10*, vacuum window; *11*, erection waveguide; *12*, pump; *13*, attenuator; *14*, 7.8dB coupler; *15*, phase shifter; *16*, waveguide adapter; *17*, zeolite pump; *18*, vacuum tubing; *19*, gauging tubes; and *20*, vacuum valve.

and reliability of rf feeder components. The service life of rf power supply components is decided mainly by their electric strength since multiple sparking usually destroys the waveguide elements. Theoretical analysis and experiments devoted to electric-strength certification of rf system components indicated that all the components, save for the 7.8 dB directional coupler, exhibit at least doubled sparking margin at 1 MW of rf power. A guaranteed electric strength of the directional coupler at 1.5 MW of rf power and ambient pressure was verified after an experimental analysis of the topography of the electric field at coupling slots and reducing its amplitude.

The simple operation and stability of the output characteristics of these linacs is a consequence of a strict stabilization of rf frequency by accelerating sections and a correct design of the section cooling system. The magnetron automatically offsets deviations in section resonance frequencies that occur as a result of beam load and variation of temperature. At the same time, the cooling system does not allow a thermal disbalance of section resonance frequencies by more than ± 0.1 MHz. These accelerators can be quickly brought to the operation regime after maintenance and scheduled checks (replacement of injector cathode, replacement of magnetron, and clearing of the vacuum window) thanks to a good sealing system, which has a small volume, and a high throughput of components. Below we shortly describe the systems of RELUS-1 which are identical with other models of this series.

For RELUS machines, we developed two base designs of electron guns with a ceramic insulator, as outlined in Table 7.5.

The injection system also includes an iron-free magnetic coil that provides the optimal conditions for beam forming and a heating transformer with oil insulation of its 50-kV windings. The rated inductance at the center of the coil is 0.02 T.

Vacuum system. The operating pressure in the evacuated part of the accelerator must be at most 10^{-4} Pa. The evacuated volume of RELUS-1 is formed by an electron gun evacuated through the buncher, buncher itself, two accelerating sections, waveguide bridge, waveguide absorbing load, waveguide bend, two waveguide-vacuum manifolds, bellows adapter, high-voltage valve, and two vacuum windows. The required design evacuation rate is 82 dm³/s enabled us to use a mass produced electric spark pump with an evacuation rate of 100 dm³/s. A pre-evacuation is provided by two zeolite facilities connected to the high-vacuum valve via a tubing. After start of the electric spark pump, the zeolite facilities are dismantled. The level of vacuum is measured by a vacuum meter. Vacuum sealing between the said units is provided by 18 copper and 4 indium washers. This design involves more than 100 brazed and welded joints, 49 of which are in the accelerating tank. The quality of these points is rather high.

Table 7.5 Characteristics of RELUS electron guns

	Option 1	Option 2
Injection voltage	40–45 kV	40–45 kV
Injection current	0–0.4 A	0–0.4 A
Beam radius at given coordinate	0.85	0.85
Beam divergence, deg.	1.3	0.45
Cathode	Extruded aluminum cathode with spherical surface and indirect heating	Flat, two-filament, spiral, tungsten, direct-heating cathode
Rated current of heating	4 A	14.5 A
Rated voltage of heating	13 V	11.3 V

Note: The beam radius and its divergence angle are given at the middle of the fist cell of the buncher at a distance of 72 mm from the beam outlet.

After the final assembly of the accelerator, the working pressure of 5×10^{-4} Pa has been achieved in 6 hours of evacuation with a high-vacuum pump. After a two-month break in pumping the pressure never rose above 3×10^{-3} Pa, therefore, one could immediately use the high-vacuum pump to obtain the operating pressure in 45 minutes. When the accelerator was used every day, one needed only fifteen minutes of pump operation to reach the operating mode.

Focusing system. This system realizes a magnetic field on the accelerator axis such that completely driven bunches entrapped in the accelerating mode. It is made in the form of three coils of 250 turns of 100×0.1-mm oxidized aluminum foil. The magnetic field attains a maximum around 10^5 A/m at supply current of $I_{fs} = 32$ A. The magnetic lens of the injector is powered by an individual source. The rated inductance at the axis is 0.02 T is achieved at a current of 0.3 A.

Cooling system. This system has two main pipes: inlet and drain with a copper tubing to and from cooled systems. Water is pumped to the magnetron, buncher, 1st and 2nd accelerating tanks, two absorbing loads, focusing coils, attenuator, and vacuum pump. Thus, between two main pipes, this system establishes a network of parallel cooling systems, each of which is equipped with a valve and a hydraulic interlock. The thermal power released in the three-tank accelerator does not exceed 1 kcal/s, thus, at a discharge of 0.25 kg/s, the cooling fluid temperature increases by 4°C. The flow rate of water in the parallel cooling networks of accelerating tanks is controlled so that, at the rated power, their temperatures differ at most by 1°C. As temperature probes, we use thermoresistors fixed at the inlets of power to the accelerating tanks.

In order to study the parameters of accelerated beam, RELUS linacs were equipped with a magnetic spectrometer (data plotted on a recorder) and a plate probe of average energy. The pulse current of the beam was determined from the oscillogram of voltage across the load resistances of a Faraday cylinder and absorbing plate. The error of spectrometric measurements was $\Delta W_k/W_k \leq 3.3\%$, and the error of average energy measurements did not exceed $\Delta \overline{W}_k/\overline{W}_k \leq 6\%$.

The initial stage of accelerator startup included training of rf assemblies and a study of the frequency stability of the rf generator. After these procedures we attained stable operation at a magnetron power up to 1.5 MW, measured the frequency stabilization coefficient $k_{st} = 12\pm2$, and the band of capture into a stable single-frequency regime $\Delta f_{cap} = 1.8$ MHz. The rf generator power was estimated with an error $\Delta P_{rf}/P_{rf} \leq 5\%$ from the anode current of the magnetron with a calibration plot obtained for a standard absorbing load connected with a calorimeter.

A typical profile of oscillograms taken at output facilities is presented in Fig 7.8. The envelope of an rf pulse from the first accelerating tank indicates the transient process as the tank is being filled with electromagnetic energy. The filling time $t_f = 0.8$ μs corresponds to a loaded tank $Q_l = 4.4 \times 10^4$. The leak resistances of the Faraday cylinder and absorbing plate is $r_{leak} = 75 \, \Omega$, and the voltages across them are 3 V. The beam current summing the leakage Faraday current and absorbing plate

Figure 7.8. Oscillograms taken from output facilities of RELUS-1 accelerator: *1*, envelope of an rf pulse in the first accelerating tank; *2*, beam current pulse from the Faraday cylinder; and *3*, beam current pulse from the plate probe. One vertical division is equal to 1 V; one horizontal division amounts to 0.5 μs.

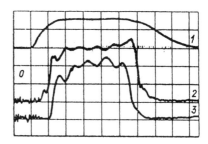

current is $I = I_{Fc} + I_{pl} = 80$ mA. The thickness of the absorbing aluminum plate is $x_{pl} = 1.08$ g/cm². The current passage coefficient $\eta = I_{Fc}/(I_{pl} + I_{Fc}) = 0.5$ which corresponds to the average (kinetic) energy of electrons $\overline{W}_k = 3.5$ MeV.

The loading characteristics of RELUS-1 for different rf powers and optimal focusing and phasing parameters are presented in Fig. 7.9. At $I = 5$ mA and $P_{rf} = 1.2$ MW, the kinetic energy of accelerated electrons was as high as $\overline{W}_k = 5.4$ MeV. At point *A*, the pulse current power is 360 kW, and the electron efficiency of the machine reaches 30%. We also present here the loading characteristics calculated with the LUCH-24 program. As can be seen, the calculated and measured data coincide within a certain error.

The results of an analysis of accelerator operation in different beam focusing modes are presented n Fig. 7.10. The maximum current transfer is achieved at $I_{fs} = 16$ A, which corresponds to a

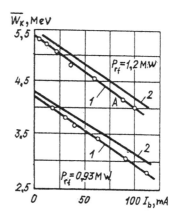

Figure 7.9. (1) Measured and (2) calculated RELUS-1 loading characteristics for optimum focusing and phasing.

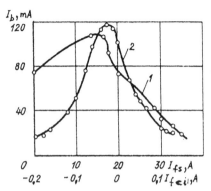

Figure 7.10. Beam current in RELUS-1 versus (1) current in the focusing coils I_{fs} (2) current in the injector focusing coils I_{fci} and $I_{fs} = 16$ A.

focusing magnetic field of 6×10^4 A/m and a current in the winding of the focusing injector coil $I_{fci} = 25$ mA. The focusing coil of the injector offsets the magnetic field of the focusing system near the cathode. The maximum lock coefficient $k_{lock,max} = 32\%$ corresponds to the calculated value. Data presented in Fig. 7.10 demonstrate that the linac can operate with a standing wave in the sine-reversible rf focusing mode without an external magnetic field. At $I_{fs} = 0$, the lock coefficient $k_{lock} = 20\%$ corresponds to theoretical estimates.

Figure 7.11 shows the dependence of the average beam kinetic energy on the position of phase shifter in the buncher feed circuit, which is equivalent to the variation of the phase of bunches at the inlet to the first accelerating tank. The shifter position $\varphi = 130°$ approximately corresponds to the location of bunches at the center of the first accelerating gap with a maximum of the accelerating electric field. It is worth noting that the variation of beam energy

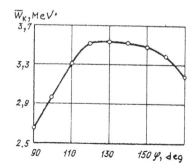

Figure 7.11. Average beam kinetic energy in RELUS-1 as a function of the phase shifter position in the buncher feed circuit ($P_{rf} = 850$ kW, $P_{bunch} = 135$ kW, $I_b = 35$ mA, and $P_{tank} = 570$ kW).

Figure 7.12. Dependence of beam average energy and current on the level of rf power fed to accelerating tanks in RELUS-1 (P_{rf} = 850 kW and P_{bunch} = 135 kW).

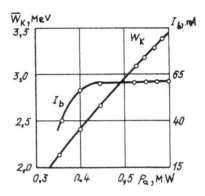

by varying the inlet phase is accompanied by a variation of the lock coefficient which reduces to a few percent when the inlet phase deviates by ±50° from the optimal level.

The output energy of the beam can be varied by changing with an attenuator the rf power fed to accelerating tanks at P_{bunch} = constant. Figure 7.12 indicates that intensity can be maintained constant with a variation depth of about 40%.

Figure 7.13 shows the dependence of beam energy and current on the magnetron power. Here, the lock coefficient also varies in parallel with the power, however, not so rapidly as for the inlet phase. In this case, beam parameter variation is more energy efficient than in the case of power takeoff by an attenuator. Therefore, this type of control should be used in operation.

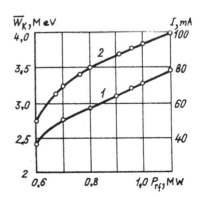

Figure 7.13. Dependencies of (1) average energy and (2) beam current on generator power fed to RELUS-1 (I_{fs} = 16 A and I_{fci} = 0).

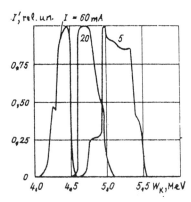

Figure 7.14. Energy spectrum of an electron beam for different beam currents in RELUS-1 (P_{rf} = 1.2 MW).

Figures 7.14–7.16 present the spectral characteristics of RELUS-1 copied from the recorder of a magnetic analyzer. Figure 7.14 illustrates the spectral evolution with the beam load at an rf power P_{rf} = 1.2 MW. Figure 7.15 presents the spectral characteristics for different powers of the magnetron, and Fig. 7.16 presents the characteristics for different phases of inlet to the center of the first accelerating gap. For $P_{rf} \geq 0.9$ MW and the rated beam current I = 50 mA, the relative spectral width at half maximum varies from 8.5 to 4.5%.

The transverse beam size was estimated from its footprint on a photographic paper placed in the output window. The prints obtained under the optimal focusing conditions distinctly reveal

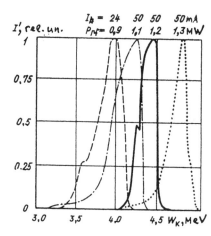

Figure 7.15. Energy spectrum of an electron beam for different rf powers in RELUS-1.

Figure 7.16. Beam energy spectrum as a function of the inlet phase of bunches in accelerating tanks of RELUS-1 ($P_{rf} = 1.1$ MW).

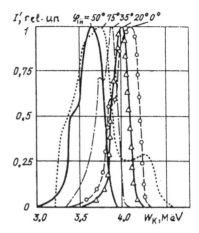

three parts: a most intensive core with a radius of 1.2 mm, a 3-mm halo, and a 4.5-mm outer halo. The core contains 90% of the intensity. When the accelerator operates in the rf focusing mode ($I_{fs} = 0$), the core radius increases up to 2.5 mm.

Since the accelerator was commissioned in 1982, we studied its characteristics, operating regimes, and reliability of its systems in long operating sessions. The utilization factor in training or

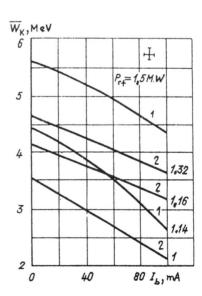

Figure 7.17. RELUS-2: (*1*) calculated and (*2*) experimental load characteristics.

research programs, defined as the time of working with the beam
relative to the accelerator operating time, has been increasing
with each year as 60% in 1982, 64% in 1983, 65% in 1984, and
66% in 1985.

Figure 7.17 presents the experimental and load characteristics
of RELUS-2 for rf powers of 1, 1.16, and 1.32 MW. At
P_{rf} = 1.14 MW, the theoretical curve coincides with experimental
data accurate to within ±10% over the whole range of accelerated
intensities up to 0.1 A.

Figure 7.18a presents the spectrum of accelerated electrons
for a magnetron power of P_{rf} = 1.4 MW and I = 14 mA. At this

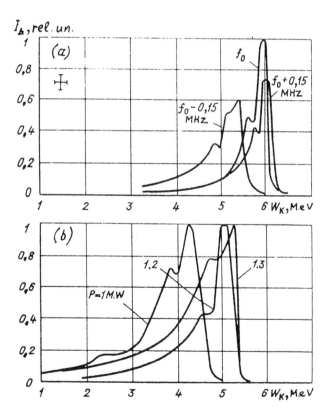

Figure 7.18. Energy spectrum of a beam versus (a) rf
generator frequency and (b) rf generator power fed to
RELUS-2.

current, we achieved the maximum kinetic energy of the beam $W_k = 6$ MeV with an energy spectrum width $\Delta W_k/W_k \leq 5.5\%$ at half maximum. Figure 7.18b indicates that lower rf powers degrade the bunching parameters and, at $P_{rf} = 1$ MW, the spectrum width increased by 25%. These data correspond to calculation of longitudinal dynamics and are typical of specialized setups. Figure 7.18a demonstrates that the stability of output beam parameters is mainly defined by the generator frequency deviation. This fact was proved by a series of measurements performed every two minutes after the accelerator has reached a stable thermal regime. These measurements indicated that the instability of energy at the spectral maximum did not exceed ±3.5% during 15 min and was associated predominantly with the oscillations of the magnetron anode voltage, which, during the same time span, reach ±3%.

For the RELUS-2 accelerator, we studied the dependence of beam current and energy on the injection voltage. In this experiment, the injector was fed from an independent modulator. Results presented in Fig. 7.19 prove that, in agreement with calculations, 40 kV is an optimal injection voltage.

The dependence of beam current and average energy on the external focusing magnetic field is presented in Fig. 7.20. Here, the current $I_{fs} = 60$ A corresponds to the on-axis inductance 0.2 T and the lock coefficient $k_{lock} \sim 1$. The rated inductance of the focusing field (0.12 T at $I_{fs} = 32$ A) corresponds to a lock coefficient of 50%. In the absence of a focusing magnetic field, the lock

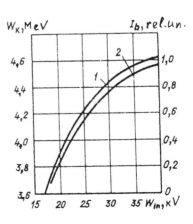

Figure 7.19. Dependencies of (1) kinetic energy at the spectral maximum and (2) beam current on the injection voltage for RELUS-2 ($P_{rf} = 1.1$ MW).

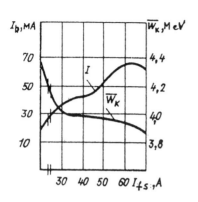

Figure 7.20. Beam characteristics of RELUS-2 for different focusing conditions ($P_{rf} = 1.3$ MW).

coefficient reduces to 20%, however, at this level, the accelerator operates in a rated regime with $I = 50$ mA, since the maximum injection current $I_{inj,max} = 375$ mA. These experimental data have proved the correctness of radial dynamics calculations with the Radius computer code.

The spectral characteristics of SW linacs can be substantially improved if the injection pulse can be delayed relative to the rf power pulse by eliminating the acceleration during transient filling of accelerating tanks with rf power [153]. Results of a relevant investigation are presented in Fig. 7.21. Spectrum *1* relates to the regime when the injection pulse is not delayed. Here, $\Delta W_k/W_k = 7.3\%$ and $W_{max} = 5.5$ MeV at $I = 50$ mA and rf power $P_{rf} = 1.5$ MW. With the optimal delay of injection pulse $\tau_d = 1$ µs, the energy spectrum narrows down to 4.5%, and the

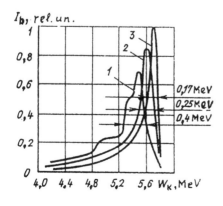

Figure 7.21. Evolution of the energy spectrum of a beam in a RELUS-2 accelerator upon a delay of the injection current pulse ($P_{rf} = 1.5$ MW and $I_b = 50$ mA): (*1*) $\tau_d = 0$, $\tau_p = 3$ µs; (*2*) $\tau_d = 1$ µs, $\tau_p = 3$ µs; and (*3*) $\tau_d = 1$ µs, $\tau_p = 1.5$ µs.

maximum energy builds up to 5.65 MeV due to a reduced beam load from the low-energy part of the spectrum. A further improvement of the spectral composition of the beam can be achieved by narrowing the delayed injection pulse down to $\tau_p =$ 1.5 μs, thus eliminating the acceleration at the trailing edge of the transient process. In the last case, $\Delta W_k/W_k = 3\%$ and the low-energy skirt of the spectrum breaks at $W_k = 4.6$ MeV. Thus, the quality of the energy spectrum of SW linacs is nowhere inferior to that of TW units.

Measurements of photographic beam footprints in RELUS-2 indicated that, at $I_{fs} = 32$ A, the core diameter is 4 mm and increases up to 7 mm in the rf focusing mode.

7.5 Uragan-2 Proton Linac

A small-size linear accelerator of protons named Uragan-2 rated for 0.5 MeV has been developed for research in electronic technology. This machine is a resonant accelerator operating at a frequency of 150 MHz. Its accelerating structure is an interdigital H-type cavity excited in the π mode.

The parameters of the accelerating structure with rf field focusing were selected with a method based on a harmonic representation of the rf field in the region of interaction. Calculations were conducted with a model of two harmonics one of which varies in phase with the beam and provides acceleration of protons, whereas the other is asynchronous and used for beam focusing.

In order to increase beam intensity, we decided to do without bunching in the initial part of the structure which is typical of accelerators with phase variable focusing. Instead we smoothly decreased the phase length of a bunch as the particle energy increased. In order to conserve the adiabatic invariant of longitudinal motion for an accelerating beam, the amplitude of the synchronous harmonic must increase with energy [47].

Using these dependencies and equations for average motion of particles in an accelerator with rf focusing, we selected functions of amplitude and velocity of the asynchronous harmonic such that provide the desired focusing conditions. In calculations, we assumed that the minimal (over all bunch particles) phase advance of the transverse oscillations over a length $\beta_{ph}\lambda$ remained invariable and equal to 20° along the entire structure. Using these functions of amplitudes and velocities of harmonics, we determined the field over half periods of the rf structure which consisted of two drift half-tubes and an accelerating gap.

It is worth noting that the selection of amplitude and velocity functions of the asynchronous harmonic is far from unique and is limited by a number of conditions. The most important condition is to provide the required profile of the potential difference across gaps over the accelerator length. This profile is determined by the choice of an accelerating structure and capabilities of tuning elements.

Figure 7.22 depicts an accelerating cavity of the linac. A specific feature of this design is a container assembly which can be removed from the shell. The assembly is a rectangular frame

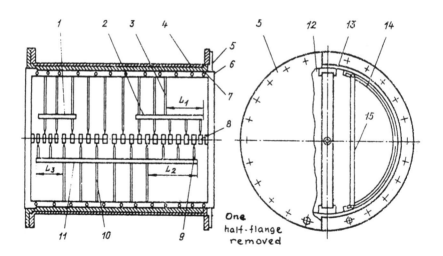

Figure 7.22 Accelerating cavity of the Uragan-2 linac: *1* and *2,* short-circuit ties; *3,* movable short-circuit rod; *4,* shell; *5,* end walls; *6,* container assembly; *7,* container guides; *8,* drift tubes; *9* and *10,* short and long stems of drift tubes; *11,* short circuit tie; *12,* bolts of clamping planks; *13,* planks clamping the container to the guides; and *14* arc to mount the tuning bar *15.*

brazed of 24×20-mm copper bars. This frame carries long stems of drift tubes fixed to it by brazing. Drift tubes are connected by brazing to short circuit ties made as 20×20-mm copper bars. Short stems of drift tubes (10-mm diameter rods) are brazed to the ends of these ties. The container is fixed to the guides, brazed to the shell walls, by two special clamping planks. Clamping is effected by bolts. End walls and arcs for mounting the tuning plungers are attached to the shell by bolts. All the components are made of oxide-free copper. They are brazed by 72% silver solder in furnaces under hydrogen. The brazed structure was artificially aged and regained the initial stiffness after a week.

In order to achieve a high Q-factor, the surface roughness of current currying surfaces was made below $0.32\ \mu m$. After mechanical treatment, all details of the accelerating structure were subjected to electropolishing.

The removable container design considerably simplifies the manufacturing technology of such accelerating cavity (fine adjustment of beam channel, brazing, and electropolishing) to say nothing of the replacement of the container.

Tuneup of accelerating structure was effected as follows [146]. First, we studied the effect of each end resonance element on the electric field distribution. Figure 7.23 displays the dependence of the maximum electric field in gaps of drift tubes on the number of a gap for three values of L_1 and fixed values of L_2 and L_3. This profile comes close of all to the calculated distribution of the field in gaps for $L_1 = 192$ mm, $L_2 = 140$ mm, and $L_3 = 40$ mm. However, owing to a strong coupling between separate rods we could not realize the desired nonuniformity of potential difference along the cavity length. Therefore, with an eye to the capacity of tuning, we calculated a structure with a smoothed dependence of potential difference over the cavity length, i.e., at the said values of tuning elements. This calculation produced satisfactory results. In calculating the dynamics, the choice of the first gap provides the best dynamic matching of the phase beam volume with the rf accelerating-focusing channel.

Since the resonant frequencies of the accelerating cavity vary substantially in tuneup, we used tuning bars to compensate for

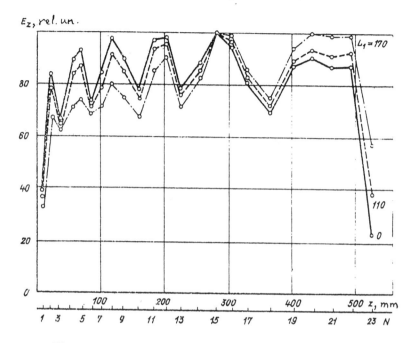

Figure 7.23. Dependence of the peak strength of the electric field in a gap on the gap number for $L_1 = 0$, 110, and 170 mm, $L_2 = 170$ mm, and $L_3 = 40$ mm.

these variations. These ties are conducting rods (*15* in Fig. 7.22) placed in the middle of the cavity parallel to the drift tube stems and electrically connected to the body of the cavity at their ends. The shell arc φ contracted by such a rod, is defined by [128]

$$f_0 - f_1 \approx \frac{\mu_0 R^3 \left[\pi f_1(\varphi - \sin\varphi)\xi_{\mathrm{M}}\right]^2}{8\left(\sin\dfrac{\varphi}{2} + \dfrac{\varphi}{2}\right) \ln\left[\dfrac{R}{\pi a}\left(\sin\dfrac{\varphi}{2} + \dfrac{\varphi}{2}\right)\right]}, \tag{7.10}$$

where f_1 is the initial resonance frequency, f_0 is the given operating frequency, R is the cavity radius, a is the rod radius, and $\xi_M = H_z/(PQ)^{1/2}$ is the parameter of the magnetic field on the surface of the body in the middle of arc φ.

After tuning of the accelerating cavity for the given distribution of the electric field in gaps, its main electrodynamic characteristics were as follows: Q-factor 4800 and shunt impedance 70 MΩ or 130 MΩ/m. At 35 kW of the rated rf power, the maximum field strength, achieved in gap 16, amounts to 7.6 MV/m. The rf input coupler is made as a coaxial line with a loop at the end. In order to provide for a critical coupling of the rf generator with the cavity, the loop was made of 15×8-mm plates and had an area of 4×10^{-4} m^2. The loop inductance amounted to 5×10^{-9} H.

The block diagram o the Uragan-2 accelerator is presented in Fig. 7.24. The main characteristics of this machine are listed below.

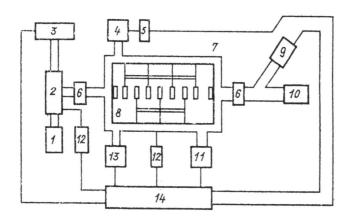

Figure 7.24. Block scheme of the Uragan-2 accelerator: *1*, vacuum system of the injector; *2*, injector; *3*, injector power supply; *4*, rf generator; *5*, modulator; *6*, separator; *7*, vacuum sealed casing; *8*, accelerating structure; *9*, beam measurement system; *10*, operating chamber; *11*, zeolite unit; *12*, vacuum-meters; *13*, high-vacuum pump; and *14*, control board.

Injection energy	27.5 keV
Output energy	500 keV
Average output current	8 μA
Current pulse length	100 μs
RF generator frequency	150 MHz
Rated rf power of the source	35 kW
Energy instability	5%
Length of accelerating cavity	600 mm
Number of accelerating gaps	23
Drift tube aperture diameter	10 mm
Maximum strength at gap center	80 kV/cm
Maximum field strength on the surface of drift tubes	200 kV/cm

Protons are injected into the cavity with a dua-plasmotron and a system of electrostatic lenses. The main parameters of the beam at the inlet into the accelerating cavity are as follows: beam energy 30 keV, proton current 20 mA, pulse length 100 μs, beam diameter 5 mm, energy spread 5%, and emittance 40 mrad/cm. The injector is provided with an individual power supply system.

The vacuum system of the machine is capable of maintaining a pressure not higher than 10^{-4} Pa in the sample chamber and in the vacuum chamber of the accelerating structure, and a pressure not higher than 10^{-3} Pa in the vacuum chamber of the injector.

The rf generator in this system is a three-stage pulse amplifier. The output stage is built around a triode connected into a grounded grid circuit. A mass produced generator is used as a master oscillator. The high-voltage modulator is made on the basis of a nonuniform forming line with a partial sparking and provides 14 kV of pulse voltage with a length of 100 μs to the anode of the triode. The generator is matched with the accelerating cavity via a telescoping coaxial line. The output power at measured 150 MHz achieves 50 kW. The generator frequency can be retuned within 6%.

Notation

a	beam channel and disk aperture radius
b	internal radius of DLWs and BPSs
$\mathbf{B}$	magnetic inductance
d	axial cell size
d_g	accelerating gap length
D	period of accelerating structure
$\mathbf{D}$	electric displacement
$\mathbf{E}, E_r, E_\varphi, E_z$	electric field strength and its components
E_{zm}	axial amplitude of the mth harmonic
E_0	axial amplitude of the fundamental harmonic
$E_{s,max}$	maximum field strength on structure surface
$\mathbf{E}^v (\mathbf{H}^v)$	vortex part of electric (magnetic) field
$\mathbf{E}^p (\mathbf{H}^p)$	potential part of electric (magnetic) field
$\mathbf{E}_m (\mathbf{H}_m)$	vortex eigenfunction of electric (magnetic) field of the cavity
$\mathbf{E}_g (\mathbf{H}_g)$	potential eigenfunction of electric (magnetic) field of the cavity
f_0	resonance frequency
$\mathbf{H}, H_r, H_\varphi, H_z$	magnetic field strength and its components
I_m, I_g, V_m, V_g	expansion coefficients on cavity eigenfunctions
J_{ext}	external current density
k	wave number in free space
k_z	longitudinal component of waveguide wavevector
k_t	transverse component of waveguide wavevector
k_c	coupling coefficient in accelerating structure
k_g	gap efficiency coefficient
k_q	structure quality coefficient
k_{ov}	overvoltage factor
L	cavity length
L_a, L_c	lengths of accelerating and coupling cells
m	number of azimuthal variations, number of spatial harmonic
n	mode index ($n = \theta/\pi$), number of variations over the radius
N	number of cells in the cavity
p	ordered number of a cell, number of longitudinal variations of the field

P	rf power flux, rf power loss in the cavity
q_i	any dimension of accelerating structure
Q_m	Q-factor of the cavity in mode m
r_{sh}	shunt impedance per unit length
$r_{sh,eff}$	effective shunt impedance per unit length
R_{sh}	shunt impedance
$R_{sh,eff}$	effective shunt impedance
R	radius of the body of the cavity
R_r	radius of sharp edge rounding
S_p	surface area of cell p
t	disk or iris thickness
T	transit time factor
U	potential difference
V_p	volume of cell p
W_{max}	maximum energy stored in the cavity
W_k	kinetic energy of particles
Z_w	wave impedance of transmission line
Z_0	wave resistance of free space
α	attenuation of rf field
β_{ph}	phase velocity in light speed units
β_{gr}	group velocity in light speed units
β_0	coefficient of coupling of accelerating structure and rf feeding waveguide
ε	relative dielectric permittivity of the medium
ε_0	electric constant
ε_a	absolute dielectric permittivity of the medium
θ	mode (phase advance of the em field per cell)
λ	wavelength in free space
λ_{wg}	wavelength in waveguide
μ	relative magnetic permeability of the medium
μ_a	absolute magnetic permeability of the medium
μ_0	magnetic constant
ρ	charge bulk density
σ_e	specific conductance of the medium

References

1. Val'dner, O.A., Vlasov, A.D., and Shal'nov, A.V., *Lineinye uskoriteli* (Linear Accelerators), Moscow: Atomizdat, 1969.

2. Val'dner, O.A., Shal'nov, A.V., and Didenko, A.N., *Uskoryayushchie volnovody* (Accelerating Waveguides), Moscow: Atomizdat, 1973.

3. Val'dner, O.A., Sobenin, N.P., Zverev, B.V., and Shchedrin, I.S., (Eds.), *Spravochnik to diafragmirovannym volnovodam* (Handbook of Disk-Loaded Waveguides), Moscow: Atomizdat, 1968, 1978.

4. Val'dner, O.A., Sobenin, N.P., Zverev, B.V., and Shchedrin, I.S. (Eds.), *Diafragmirovannye volnovody: Spravochnik* (Disk-Loaded Waveguides. A Handbook), Moscow: Energoatomizdat, 1991, 3rd ed.

5. Grigor'ev, A.D. and Yankevich, V.B., *Rezonatory i rezonatornye zamedlyayushchie sistemy SVCh* (Cavities and Resonator Decelerating Microwave Systems), Moscow: Radio i Svyaz, 1984.

6. Naidenko, V.I. and Dubrovka, F.F., *Aksial'no-simmetrichnye periodicheskie struktury i rezonatory* (Axially Symmetric Periodic Structures and Cavities), Kiev: Vishcha Shkola, 1985.

7. Murin, B.P., Bondarev, B.I., Kushchin, V.V., and Fedotov, A.P., *Lineinye uskoriteli ionov* (Ion Linacs), Moscow: Atomizdat, 1978.

8. Lebedev, A.N. and Shal'nov, A.V., *Lineinye uskoriteli* (Linear Accelerators), Moscow: Energoatomizdat, 1983.

9. Kapchinskii, I.M., *Teoriya lineinykh resonansnykh uskoritelei* (Theory of Linear Resonant Accelerators), Moscow: Energoizdat, 1982.

10. Karetnikov, D.V., Slivkov, I.N., Teplyakov, V.A., et al., *Lineinye uskoriteli ionov* (Linear Ion Accelerators), Moscow: Gosatomizdat, 1962.

11. Milovanov, O.S. and Sobenin, N.P., *Tekhnika sverkhvysokikh chastot* (Microwave Circuits and Devices), Moscow: Atomizdat, 1980.

12. Lapostolle, P.M. and Septier, A.I., *Linear Accelerators*, Amsterdam: North-Holland, 1970.

13. Weaver, J.N., *Measuring, Calculating and Estimating PEP's Parasitic Mode Loss Parameter*, SLAC-PEP-352, 1981.

14. Miller, R.H., *Comparison of standing wave and travelling-wave structures*, SLAC-PUB-3935(A). 1986.

15. Giranlt, P. and Trong, D., 4π/5 Backward TW structure tested for electron linacs optimization, *European Particle Accelerator Conference. Rome.* June 7–11, 1988, 1114–1116.

16. Trong, D., Electron linac optimization for short rf and beam pulse lengths, *IEEE Trans. Nucl. Sci.,* 1985, **NS-32**(5), 3243–3245.

17. Hoffswell, R.A. and Lszewski, R.M., Higher modes in the coupling of coaxial and annular ring coupled linac structures, *IEEE Trans. Nucl. Sci.,* 1983, **NS-30**(4), 3588–3589.

18. Andreev, V.G., Geometry of a structure with sign-reversing field excited in the π/2 mode, *Zh. Tekh. Fiz.,* 1971, **41**, 788–796.

19. Tanaka, T, Hayakawa, K., Tsukada, K., et al., Test of a 1-m long disk-and-washer accelerating tube, *Proc. of the 1984 Linear Accelerating Conference,* Darmstadt, GSI-84-11, 1984, 229–231.

20. Wilson, P.B., RF-driven linear colliders, *SLAC-PUB*-4310, 1987.

21. Balakin, V.E. and Srkinskii, A.N., VLEPP project status, *Trudy XIII Mezhd. konf. po chastitsam vysokikh energii* (Proc. XIII Int. Conf. on High Energy Particles), Novosibirsk: Nauka, 1987, **1**, 101–108 (in Russian).

22. Balakin, V.E. and Novokhitskii, A.V., Beam dynamics in a linear accelerator VLEPP, *Trudy XIII Mezhd. konf. po chastitsam vysokikh energii* (Proc. XIII Int. Conf. on High Energy Particles), Novosibirsk: Nauka, 1987, **1**, 146–150 (in Russian).

23. Schnell, W., *Liner Collider Studies in Europe,* CERN, LEP-RF, 88-30, 1988.

24. Wilson, I., Schnell, W, and Henke, H., *Design and Fabrication Studies of High Gradient Accelerating Structures,* CERN, LEP-RF, 88-50, 1988.

25. Loev, G.A., *Some Issues Involved in Designing a 1 TeV (c.m.) $e^\pm$ Linear Collider Using Conventional Technology,* SLAC-PUB-3892, 1986.

26. Boiteux, J.-P., Garvey, T., Geschonke, G., et. al., *Studies of R.F. Accelerating Structure for an Electron Linear Collider,* CERN: LEP-RF, 87–25, 1987, CLIC Note 36.

27. Farkas, Z.D. and Wilson, P.B., *Comparison of High Group Velocity Accelerating Structures,* SLAC-PUB-4088, 1987.

28. Alimov, A.S., Zverev, B.V., Sandalov, A.I., et al., Linear accelerator for race-track cw microtron, *Trudy X Vsesoyuzn. soveshchaniya. po uskoritelyam zaryazhennykh chastits* (Proc. X All-Union Meeting. on Accelerators of Charged Particles), Dubna, 1987, **1**, 190–192 (in Russian).

29. Penner, S., Linacs for Microtron and Pulse Stretchers, *Linear Accelerator Conf. Proc.*, SLAC-303, 1986, 416–420.

30. Dwersteg, B., Selsseberg, E., and Zaltaghari, A., Higher order mode for normal conductivity DOROS 5-cell cavities, *IEEE Trans. Nucl. Sci.*, 1985, **NS-32**(5), 2797–2799.

31. Allen, M.A., Wilson, P., et al., RF system for the PEP storage ring, *IEEE Trans. Nucl. Sci.*, 1977, **NS-24**(3), 1780–1782.

32. Husmann, D.A., Stretcher ring and post accelerator for the Boon 2.5 GeV electron synchrotron, *IEEE Trans. Nucl. Sci.*, 1983, **NS-30**(4), 3590–3592.

33. Karnaukhov, I.M., Tonkov, Yu.P., Telegin, Yu.N., et al., Electrodynamic characteristics and choice of a structure for a stretcher ring, *Preprint Kharkov Physicotechn. Inst. 89-6*, Kharkov, 1989.

34. Fornaca, S., Hess, C.H., Schwettman, H.A., et al., Experimental investigation of low frequency modes of a single cell r.f. cavity, *Proc. of the 1987 IEEE Particles Accelerator Conf.*, IEEE Catalog no. 870-CH-2387-9, **3**, 1818–1820.

35. Schwettman, H.A., Smith, T.J., and Hess, C.H., Electron acceleration using high gradient single cell resonators, *IEEE Trans. Nucl. Sci.*, 1987, **NS-32**(5), 2927–2929.

36. Veshcherevich, V.G., Karliner, M.M., Nezhevenko, O.A., et al., High-current linac of VEPP-4 system. Accelerating Structure, *Preprint of Inst. Nucl. Phys. 83-148*, Novosibirsk, 1983.

37. Karliner, M.M., Nezhevenko, O.A., Ostirko, G.N., et al., 100 MeV electron linac with the DAW structures an injector for the Siberia-2 storage ring, *Proc. of the First European Particle Accelerator Conf.*, Rome, 1988, 602–604.

38. Karliner, M.M., Nezhevenko, O.A., Fomel', B.M., et al., Comparison of Accelerating Structures Operating on Stored Energy, *Preprint Inst. Nucl. Phys. 86-146*, Novosibirsk, 1986.

39. Wideröe, R., *Archiv für Electrotechnik*, 1928, **21**, 387–391.

40. Angert, N., Bohne, D., Klabunde, J., et al., The UNILAC upgrading program, *Proc. of the 1979 Linear Accelerator Conf.*, New York, 1979, 17–21.

41. Blewett, J.D., Linear accelerator injectors for proton synchrotrons, *Proc. Symp. on High Energy Accelerators and Pion Physics*, Geneva, 1956, **1**, 162–170.

42. Pottier, J., *One nouvelle structure a cavity resonante pour accelerateurs lineaires d'ions*, Note C.E.A., 1957, no. 195.

43. Kovpak, I.E., Baranov, L.N., and Zeidlits, P.M., Linear H-type proton accelerator for 2.5 MeV, *Ukr. Fiz. Zh.*, 1968, 13(4), 552–555.

44. Bomko, V.A., D'yachenko, AF., Pipa, A.V., et al., Tuneup of the main tank of LUMZI in the H111 mode, *Voprosy atomnoi nauki i tekhniki* (Topics in Atomic Science and Technology. Ser. Technology of Physical Experimentation), Kharkov: KhFTI, 1986, 1(27), 17–21.

45. Hirao, Y., An overview of accelerator developments in Japan, *IEEE Trans. Nucl. Sci.*, 1985, **NS-32**(5), 1565–1570.

46. Weiss, T., Klein, H., and Schempp, A., Highly efficient interdigital H-type resonator for molecular ions, *IEEE Trans. Nucl. Sci.*, 1983, **NS-30**(4), 3548–3550.

47. Bondarenko, P.B, Pronin, A.N., and Sobenin, N.P., Accelerating structure for a linear resonance proton accelerator with low injection energy, *Trudy X Vsesouzn. soveshchaniya po uskoritelyam zaryazhennykh chastits* (Proc. X All-Union Meeting on Accelerators of Charged Particles), Dubna, 1987, **1**, 243–246.

48. Mashkovtsev, B.M., Tsibirov, K.N., and Emelin, B.F., *Teoriya volnovodov* (Waveguide Theory), Moscow: Nauka, 1966.

49. Lopukhin, V.M., *Vozbuzhdenie elektromagnitnykh kolebanii i voln elektronnymi potokami* (Excitation of Electromagnetic Oscillations and Waves by Electromagnetic Fluxes), Moscow: Izd. TTL, 1953.

50. Mikhlin, S.G., *Variatsionnye metody v matematicheskoi fizike* (Variational Methods in Mathematical Physics), Moscow: Nauka, 1970.

51. Richtmeyer, R.D. and Morton K.W., *Difference Methods for Initial-Value Problems*, New York: Interscience, 1967.

52. Voitovich, N.N., Katsenelenbaum, B.Z., and Sivov, A.N., *Obobchshennyi metod sobstvennykh kolebanii v teorii diffraktsiii* (Generalized Method of Resonance Oscillations in Diffraction Theory), Moscow: Nauka, 1977.

53. Nikol'skii, V.V. and Nikol'skaya, T.I., *Elektrodinamika v rasprostranenie radiovoln* (Electrodynamics and Propagation of Radio Waves), Moscow: Nauka, 1989.

54. Marchuk, G.I., *Metody vychislitel'noi matematiki* (Methods of Numerical Analysis), Moscow: Nauka, 1980.

55. Forsythe, G., Malcolm, M, and Moler, C., *Computer Methods for Mathematical Computations*, Englewood Cliffs: Prentice Hall, 1977.

56. Strang, G. and Fix, G., *An Analysis of the Finite Element Method*, Englewood Cliffs: Prentice hall, 1973.

57. Marchuk, G.I. and Agoshkov, V.I., *Vvedenie v proektsionno-setochnye methody* (Introduction to Projection-Mesh Analysis), Moscow: Nauka, 1981.

58. Parlett, B., *The Symmetric Eigenvalue Problem*, Englewood Cliffs: Prentice Hall, 1980.

59. Rektorys, K., *Variational Methods in Mathematics, Science and Engineering*, Dordrecht: Reidel, 1983.

60. Hoit, H.S., Simmonds, D.D., and Rich, V.D., Computer analysis of 805-MHz cavities for a proton linac, *Rev. Sci. Instr.*, 1966, **37**(6), 63–70.

61. Bukharin, V.L., Novozhilov, A.E., Sobenin, N.P., et al., Modeling of Biperiodic Slow-Wave Structures, *Proc. 1979 Linear Accelerator Conference*, New York, 1979, 197–2–1.

62. Bukharin, V.L. Sobenin, N.P., and Tret'yakov, A.G., Electrodynamic characteristics of cavities in high-current accelerators, *Trudy VIII Vsesoyuznogo soveshchaniya po uskoritelyam zaryazhennykh chastits* (Proc. VIII All-Union Meeting on Accelerators of Charged Particles), Dubna, 1983, 169–171.

63. Bukharin, V.L. Sobenin, N.P., and Tret'yakov, A.G., Numerical analysis of the effect beam losses on the excitation of em fields in operating volumes of accelerators, *Teoreticheskie i eksperimental'nye issledovaniya uskoritelei zaryazhennykh chastits* (Theoretical and Experimental Studies of Accelerators of Charged Particles. MIPhI Collection of Papers), Moscow: Energoatomizdat, 1984, 76–81.

64. Bukharin, V.L. Sobenin, N.P., and Yanchenko, V.V., Electrodynamic characteristics of multiply-coupled accelerator systems, *Fizika i tekhnika lineinykh uskoritelei* (Physics and Technology of Linacs. MIPhI Collection of Papers), Moscow: Energoatomizdat, 1985, 77–80.

65. Bukharin, V.L. Sobenin, N.P., and Yanchenko, V.V., Electrodynamic modeling of resonating generator-accelerator systems, *Voprosy atomnoi nauki i tekhniki* (Topics in Atomic Science and Technology), Kharkov: KhFTI, 1986, 2(28), 52–56.

66. Bukharin, V.L. Kachinskii, I.E., and Sobenin, N.P., Numerical modeling of microwave fields excited in a multiple beam cavity, *Lineinye uskoriteli* (Linear Accelerators. MIPhI collection of Papers), Moscow: Energoatomizdat, 1987, 55–60.

67. Grigor'ev, A.D., Analysis of azimuthally nonuniform modes in axially symmetric cavities, *Radiotekh. Elektron.*, 1979, **24**(6), 1211–1213.

68. Grigor'ev, A.D. and Silaev, S.A., Calculation of the electromagnetic field of azimuthally nonuniform modes in axially symmetric cavities with an arbitrary generatrix, *Elektronika SVCh*, 1981, 2, 62–65.

280   **References**

69. Weiland, T., On the computation of resonant modes in cylindrically symmetric cavities, *Nucl. Instr. Meth.*, 1983, 216, 329–348.

70. Wilhelm, W., CAVIT and CAV3D computer programs for r.f. cavities with constant cross-section or any three-dimensional form, *Particle Accelerators*, 1982, **12**, 139–145.

71. Halbach, K. and Holsinger, R.F., SUPERFISH—a computer program for evaluation of rf cavities with cylindrical symmetry, *Particle Accelerators*, 1976, **7**, 213–222.

72. Fomel', B.M., Yakovlev, V.P.,Karliner, M.M., et al., LANS—a new computer program for rf cavities with axial symmetry, *Particle Accelerators*, 1981, **11**, 172–181.

73. Karliner, M.M., Fomel', B.M., and Yakovlev, LANS-2—a computer program for azimuthally nonuniform modes in axially symmetric cavities, *Preprint Inst. Nucl. Phys.*, 83-114, Novosibirsk, 1983.

74. Daikovskii, A.G., Potapov, Yu.I., Ryabov, A.D., et al., Computation of em fields with variation in φ in axially symmetric cavities. I. Principal relations, *Preprint Inst. Phys. High Energies*, 80-107, Serpukhov, 1980.

75. Daikovskii, A.G., Potapov, Yu.I., Ryabov, A.D., et al., Evaluation of resonant em fields in cavities, II. PRUD-M computer code for azimuthally nonuniform modes in axially symmetric cavities, *Preprint Inst. Phys. High Energies*, 80-107, Serpukhov, 1980.

76. Gluckstern, R.L. Holsinger, R.F., Halbach, R., and Minerbo, G.N., ULTRAFISH—Generalization of SUPERFISH to $m \geq 1$, *Proc. Linear Accelerator Conf.*, Santa-Fe, 1981, 243–248.

77. Kaschiev, M.S., Kaschieva, V.A., Gusev, V.V. et al., Calculation of frequency spectra for em cavities with a MULTIMODE program package, *Trudy VIII vsesoyuznogo soveshchaniya po uskoritelyam zaryazhennykh chastits* (Proc. All-Union Meeting on Accelerators of Charged Particles), Dubna, 1983, **1**, 153–156.

78. Daikovsky, A.G., Portugalov, Yu.I., and Rjabov, A.D., PRUD-W: a new code to compute and design accelerating structures, *Particle Accelerators*, 1986, **17**, 201–213.

79. Hara, M., Wada, T., Fukusawa, T., and Kikuchi, F., Three-dimensional analysis of electromagnetic field by finite element method, *IEEE Trans. Nucl. Sci.*, 1983, **NS-30**(4), 1108–1113.

80. Weiland, T., On the unique solution of Maxwellian eigenvalue problem in three dimensions, *Particle Accelerators*, 1985, **17**, 227–230.

81. Klatt, B., Krawczyk, F., Novender, W.R., et al., Three-dimensional electromagnetic CAD system for magnets, rf structures and

transient wake-field calculation, *Proc. Linear Accelerator Conf.*, SLAC-Report-303, 1986, 276–278.

82. Ebeling, F., Klatt, R., Krawczyk, F., et al., *Status and Future of the 3D MAFIA Group of Codes*, DESY M-88-15, 1988.

83. Cooper, R.K., Browman, M.J., and Weiland, T., Three-dimensional rf structure calculations, *Nucl. Instr. Meth.*, 1987, **B40/41**, 959–964.

84. Ivanov, V.D., Karliner, M.M., Teryaev, V.E., et al., Application of the method of boundary integral equations to calculation of high-frequency resonators, *Zh. Tekh. Fiz. Mat. Fiz.*, 1986, 12, 1900–1908.

85. McKrecken, D. and Dorn, W., *Numerical Methods and FORTRAN Programming with Applications in Engineering and Science*, New York, Wiley, 1965.

86. Nogle, D.E., Knapp, E.A., and Knapp, B.C., A coupled resonator model for standing wave accelerator tanks, *Rev. Sci. Instr.*, 1967, **38**(3), 1583–1587.

87. Kalyuzhnyi, V.E., Calculation of accelerating structures with different cavities and narrow slots, *Zh. Tekh. Fiz.*, 1978, **48**(6), 1228–1233.

88. Sobenin, N.P., Stepnov, V.V., and Shkol'nikov, E.Ya, Beam loading in biperiodic structures in a stationary regime, *Uskoriteli* (Accelerators. MIPhI Collection of Papers), Moscow: Atomizdat, 1979, 17, 88–93.

89. Romanov, G.V., Models of coupled oscillators to describe accelerating structures, *Preprint Inst. Nucl. Res.*, P-0563, Moscow, 1987.

90. Allen, M.A., and Kino, G.A., On the theory of the strongly coupled cavity chains, *IRE Trans. Microwave Theory Technol.*, 1960, **MTT-8**, 362–372.

91. Bevense, R., *Electromagnetic Slow Wave Systems*, New York: Wiley, 1964.

92. Bukharin, V.N., Zverev, B.V., and Yanchenko, V.V., Engineering design of biperiodic decelerating structures, *Lineinye uskoriteli* (Linear accelerators. MIPhI Collection of Papers), Moscow: Energoatomizdat, 1987, 13–18.

93. Semyonov, N.A., *Tekhnicheskaya elektrodinamika* (Technical electrodynamics), Moscow: Svyaz, 1973.

94. Altman, J.L., *Microwave Circuits*, New York: Van Nostrand, 1964.

95. Zavadtsev, A.A., Zverev, B.V., and Sobenin, N.P., Accelerating system of a 5-MeV resonant electron linac, *Uskoriteli* (Accelerators. MIPhI Collection of Papers), Moscow: Atomizdat, 1979, 27–31.

96. Vodop'yanov, F.A. and Murin, B.P., Generation of rf fields by a beam of relativistic particles, *Trudy Radiotechnicheskogo instituta AN*

SSSR (Proc. Radio Engineering Inst. Acad. Sci. USSR), Moscow: Radiotekhnicheskii Inst., 1975, 20–53.

97. Zverev, B.V. and Sobenin, N.P., Development of small-size linear electron accelerators in MIPhI, *Voprosy atomnoi nauki i tekhniki* (Topics in Atomic Science and Technology), Kharkov: KhFTI, 1986, 1(27), 3–8.

98. Shirman, A.D., *Radiovolnovody i ob"emnye rezonatory* (Radio Waveguides and Cavities), Moscow: Svyazizdat, 1959.

99. Ramo, S. and Winnery, J., *Fields and Waves in Modern Radio Engineering,*

100. Wilson, P.B., *High Energy Electron Linacs: Applications to Storage Ring R.F. Systems and Liner Colliders*, SLAC-PUB-2884, 1982.

101. Zavadtsev, A.A., Zverev, B.V., and Sobenin, N.P., Standing wave accelerating structures, *Zh. Tekh. Fiz.*, **54**(1), 82–87.

102. Balakin, B.E., Brezhnev, O.N., Zakhvatkin, M.N., et al., Investigation of the limiting acceleration rate in a linear accelerator of VLEPP, *Trudy XIII Mezhdunarodnoi konferentsii po uskoritelyam chastits vysokikh energii* (Proc. XIII Int. Conf. on Accelerators of Energetic Particles), Novosibirsk: Nauka, 1987, **1**, 144–150.

103. Schnell, W., *Radio Frequency Acceleration for Linear Colliders*, CERN-LEP-RF 86-27 and CLIC Note 24.

104. Palmer, R.B., *The Interdependence of Parameters for TeV Linear Colliders*, SLAC-PUB-4295, 1987.

105. Sobenin N.P., Ivanov S.I., and Kaluzhny, V.E.. The investigation of coupler for linear collider accelerating section. *Proc. of the Third European Particle Accelerator Conf.*, 1992, **2**, 1226–1227.

106. Deruyter, H., Hoag, H., Ko, K., and Ng, C.-K., *Symmetrical Double Input Coupler Development*, SLAC-PUB-5887, 1992, August.

107. Holtkamp, N. and Weiland, T., Structure work for an S-band linear collider, *Proc. of the XV International Conference on High Energy Accelerators*, 1992, **2**, 830–832.

108. Sobenin, N.P., Milovanov, O.S., and Kaluzhny, V.E.. Study of linear collider coupler parameters and accelerating section impedance characteristics, *Proc. of the Fourth European Particle Accelerator Conference*, 1994, 2028–2030.

109. Kaljuzhny, V.E., Sobenin, N.P., Milovanov, O.S., Equivalent scheme and parameters of disk loaded waveguide at dipole mode, *Proc. of*

the fifth European Particle Accelerator Conference (EPAC '96), Sitges (Barcelona). 1996, **3**, 2044–2046.

110. Sobenin N.P., Kaljuzhny V.E., Holtkamp N., HOM damping in SBLC accelerator section using input coupler, Proc. of the 18th International Linac Conference (LINAC '96), Geneva, 1996, **2**, 824–826.

111. Labrie, J.-P, Chan, K.C.D., McKeown, Y. et al., High-power r.f. efficient L-band linac structure, Proc. of the 1988 Linear Accelerator Conf., CEBAF, 1988, 284–290.

112. Sobenin, N.P., Zavadzev, A.A., and Petrov, A.A., The standing wave electron linac accelerating structure for technology purposes, Proc. of the Fourth European Particle Acc. Conf., ed. by V. Suller and Ch. Peuti-Jean-Genvz, Singapore: World Scientific, 1994, **3**, 2073–2075.

113. Tanabe, E., Bayer, M., and Trail, M, A small diameter standing wave linear accelerator structure, IEEE Trans. Nucl Sci., **NS-32**(5), 2975-2977.

114. Sobenin, N.P., Kandrunin, V.N., Karev, A.I., Melekhin, V.N., Shvedunov, V.I., and Trower, W.P., Proc. 1995 Particle Accelerator Conf., ed by L. Gennari, IEEE, Piscataway 1996, **1**, 1827.

115. Kostin, D.V., Melekhin, V.N., Shvedunov, V.I., Sobenin, N.P., and Trower, W.P., Applications of Accelerators in Research and Industry, ed by J. Duggan, New York: American Institute of Physics, 1997, 1135.

116. Schnell, W. and Wilson, I., Microwave Quadrupole Structures for the CERN Linear Collider, CERN, LEP-RF/92-13, 1992.

117. Schnell, W., Microwave Quadrupoles for Linear Colliders, CERN, LEP-RF 87-24, 1987;

   I. Wilson and H. Henke, Transverse Focusing Strength of CLIC Slotted Iris Accelerating Structures, CLIC Note 62, 30, 5, 1988.

118. Sobenin, N.P., Didenko, A.N., Gusarov, V.N., and Glavatskich, K.W., Investigation of the electromagnetic fields in structures with axially asymmetrical Aperture in the disk, Trudy 12-toi Rossiiskoi konf. po uskoritelyam zaryazennykh chastits (Proc. of the 12th Russian Particle Accelerator Conf.), 1992, **1**, 187.

119. Burshtein, E.L. and Voskresenskii, G.V., Lineinye uskoriteli elektronov s intersivnymi puchkami (Linear electron accelerators with intense beams), Moscow: Atomizdat, 1970.

120. Enterneuer, H., Herminghous, H., and Schöler, H., Simple countermeasures against the TM_{110}-beam-blowup-mode in biperiodic structures, *Proc. of the 1984 Linear Accelerator Conf.*, Darmstadt, GSI-84-11, 1984, 394–396.

121. Labric, J.-P., Chan, K.C.D., McKeown, J., et al., Beam excited modes in linear accelerator structures, *Proc. of the 1984 Linear Accelerator Conf.*, Darmstadt, GSI-84-11, 1984, 168–170.

122. Gurov, G.G., Zverev, B.V., Sobenin, N.P., et al., Waveguides and resonator devices, *Preprint Inst. Phys. High Energies* 79-53, Moscow, 1979.

123. Zavadtsev, A.A. Zverev, B.V., and Sobenin, N.P., Experimental investigation of a waveguiding accelerating structure for a 3-TeV storage ring, *Zh. Tekh. Fiz.*, 1983, **53**(5), 1737–1741.

124. Zavadtsev, A.A. and Zverev, B.V., Accelerating structures with transverse rods for small-size electron linacs, *Voprosy atomnoi nauki i tekhniki* (Topics in Atomic Science and Technology), Kharkov: KhFTI, 1983, 2(14), 63–65.

125. Zavadtsev, A.A. and Zverev, B.V., New accelerating systems for SW electron linacs, *Pis'ma Zh. Tekh. Fiz.*, 1981, **7**(21), 1332–1335.

126. Zavadtsev, A.A., Zverev, B.V., and Sobenin, N.P., Development of a 10-MeV SW electron linac and beam circulation, *Voprosy atomnoi nauki i tekhniki* (Topics in Atomic Science and Technology), Kharkov: KhFTI, 1981, 3(9), 3–4.

127. Bomko, V.A., Kipa, A.V., Khizhnyak, N.A., et al., Design and investigation of em characteristics of modified accelerating structures for heavy ions, *Preprint KhFTI* 82-110, Kharkov, 1982.

128. Avrelin, N.V., Gorbatko, V.I., Zverev, B.V., at al. Approximate calculation of electrodynamic characteristics of accelerating structures build around H-type cavities, *Preprint MIPhI* 078-88, Moscow, 1988.

129. Smythe, W., *Static and Dynamic Electricity*, New York: Wiley, 1950.

130. Voinov, B.S., *Shirokodiapazonnye kolebatel'nye sistemy* (Wideband Oscillating Systems), Moscow: Sovetskoe Radio, 1973.

131. Hamming, R.V., Numerical Methods for Scientists and Engineers, New York: McGraw Hill, 1962.

132. Gubanov, S.I. and Kolyaskin, A.D., Calculation of electrostatic field in accelerating channels by the Monte Carlo method supported by dependent tests, *Lineinye uskoriteli* (Linear Accelerators. MIPhI Collection of Papers), Moscow: Energoatomizdat, 1987, 45–51.

133. Vilman, V.I. (ed.), *Spravochnik po raschetu i konstruirova-niyu SVCh poloskovykh linii* (Handbook of Development and Design of Microwave Strip Lines), Moscow: Radio i Svyaz, 1982.

134. Zverev, B.V., Pronin, A.N., Sobenin, N.P., et al., Accelerating structure with a controlled distribution of the electric field in the beam channel, *Tez. dokl. X Vsesouzn. seminara po lineinym uskoritelyam zaryazhennykh chastits* (Abstracts of reports on X All-union workshop on particle linacs), Kharkov, KhFTI, 1987, 10.

135. Zverev, B.V., Ponomarenko, A.G., Sobenin, N.P., et al., Automatic test bench for cavity parameter measurement, *Trudy III Vsesoyuzn. soveshchaniya po uskoritelyam zaryazhennykh chastits* (Proc. III All-Union Workshop on particle accelerators), Moscow: Nauka, 1973, **1**, 254–257.

136. Zavadtsev, A.A., Zverev, B.V., Ruzing, V.V., et al., Series of automatic test setups for investigation of cavities and resonator systems in the range 120–2400 MHz, *Pis'ma Zh. Tekh. Fiz.*, 1981, **7**(10), 1004–1007.

137. Zverev, B.V., Sobenin, N.P., and Shchedrin, I.S., Parameterization of the dispersion curve for a disk-loaded waveguide, *Uskoriteli* (Accelerators, MIPhI Collection of Papers), Moscow: Atomizdat, 1962, 4, 7–23.

138. Sobenin, N.P., Tragov, A.G., and Shchedrin, N.S., Evaluation of attenuation in disk-loaded waveguides, *Uskoriteli* (Accelerators, MEPI Collection of Papers), Moscow: Atomizdat, 1964, 4, 6.

139. Bychkov, S.I., Burenin, N.I., and Safarov, R.T., *Stabilizatsiya chastoty generatorov SVCh* (Stabilization of Frequency of Microwave Generators), Moscow: Sovetskoe Radio, 1962.

140. Zavadtsev, A.A., Zverev, B.V., and Sobenin, N.P., Measurement of shunt impedances of cavities and slow-wave systems, *Prib. Tekh. Eksp.*, 1984, 2, 13–15.

141. Zavadtsev, A.A., Zverev, B.V., Nechaev, N.N., et al., Accelerating system of a 5-MeV electron linac, *Uskoriteli* (Accelerators, MEPI Collection of Papers), Moscow: Atomizdat, 1977, 17, 27–33.

142. Bondarenko, P.V., Pronin, A.N., Sobenin, N.P., et al., Project of small-size radiation-safe linear ion accelerator for applications, *Lineinye uskoriteli* (Linear Accelerators. MEPI Collection of Papers), Moscow: Energoatomizdat, 1987, 10–12.

143. Vikulov, V.F. and Kalyuzhnyi, V.E., Effect of fabrication inaccuracies on the performance of SW accelerating structures, *Zh. Tekh. Fiz.*, 1980, **50**(4), 773–779.

144. Zverev, B.V., Ruzin, V.V., and Sobenin, N.P., Compact electron linac with an accelerating system having annular coupling cells, *Voprosy atomnoi nauki i tekhniki* (Topics in Atomic Science and Technology), Kharkov: KhFTI, 1981, 3(9), 16–18.

145. Sobenin, N.P., Small-size SW electron linacs, *Teoreticheskie i eksperimental'nye issledovaniya uskoritlei zaryazhennykh chastits* (Theoretical and Experimental Investigations of Particles Accelerators), Moscow: Energoatomizdat, 1985, 9–14.

146. Danilov, V.D., Pronin, A.N., Sobenin, N.P., et al., Design and tuning of the accelerating structure of a liner resonance proton accelerator with low injection energy, *Voprosy atomnoi nauki i tekhniki* (Topics in Atomic Science and Technology), Kharkov: KhFTI, 1988, 1(36), 26–29.

147. Shoffstall, D.R. and Gallagher, W.J., On the relative merits of travelling-wave and resonant operation of linacs, *IEEE Trans. Nucl. Sci.*, 1985, **NS-32**(5), 3169–3171.

148. Vikulov, B.F., Zverev, B.V., Sobenin, N.P. et al., A compact SW electron linac with bridge drive system, *Uskoriteli* (Accelerators, MIPhI Collection of Papers), Moscow, Atomizdat, 1980, 19, 5–11.

149. Vikulov, B.F., Zverev, B.V., Sobenin, N.P. et al., A compact standing-wave electron linac with rf drive system using 3dB hybrid junction, *IEEE Trans. Nucl. Sci.*, 1979, **NS-26**(3), 4292–4293.

150. Zavadtsev, A.A., Kalyuzhnyi, V.E., Kaminskii, V.I., and Milovidov, O.S., Experimental study of the system of high-frequency drive of a three-tank standing-wave electron linac, *Voprosy atomnoi nauki i tekhniki* (Topics in Atomic Science and Technology), Kharkov: KhFTI, 1979, 1(3), 68–71.

151. Kaminskii, V.I. and Milovanov, O.S., Use of ferrite isolators in drive circuits for accelerating sections, *Uskoriteli* (Accelerators, MIPhI Collection of Papers), Moscow, Atomizdat, 1980, 19, 34–39.

152. Filatov, A.N. and Shilov, V.K., Changeable bunchers of RELUS-3 accelerator, *Uskoriteli* (Accelerators, MEPI Collection of Papers), Moscow, Energoatomizdat, 1983, 43–46.

153. Vikulov, V.F., Zavorotylo, V.N., Ruzin, V.V., et al., Improving the energy spectrum in standing wave accelerators by delaying injection, *Zh. Tekh, Fiz., 1981,* **52**(11), 45–47.

154. Parasitic modes removal our of disk-and-washer accelerating structure, *IEEE trans. Nucl. Sci.*, 1983, **NS-30**(4), 3575–3580.

Index

A

accelerating cavities
  electrodynamic
    measurements of, 187
accelerating cell, 248
accelerators
  Alvarez, 19
  buncher, 231
  ion, 19
amplifier, 195
attenuation constant, 98
attenuator, 189
average power, 104

B

bead-pulling technique, 189,
  193
beam break-up, 9
Biperiodic accelerating
  structures, 9, 13, 118
  rectangular, 125
  with annular ring coupling
    cells, 13, 124
  with coaxial coupling cells,
    13, 134
  with coaxial radial coupling
    cell, 133
  with on-axis coupling cells,
    13, 69, 124
boundary conditions, 28
  Shchukin-Rykov-Leontovich,
    44
BPS. *See* biperiodic structure
BPS characteristics, 119
BPSs with internal coaxial
  coupling cells, 135
bridge coupling coefficient, 241
BSCB *see* biperiodic structure
  with cross bars
buncher, 251

C

cavities
  for various oscillation modes,
    37
  with cyhlindrical symmetry
    for azimuth-dependent
    modes, 37
  with cylindrical symemtry for
    azimuth-independent
    modes, 37
cavity
  H-type, 21, 153
  H-type axially symmetric,
    169, 171
  uniform H-type, 153
  Ω-shaped, 64, 153
cavity-analogue, 64, 83
cavity-generator coupling factor,
  6
CC. *See* coupled circuits
chain of coupled circuits, 52
characteristic
  dispersion, 2
characteristics
  BPS, 119
  electrodynamic, 132
circuit
  equivalent, 52
coaxial coupling cells, 14
coefficient
  coupling, 2, 52, 73, 119, 120,
    126
  optimum coupling, 121
  overvoltage, 2, 5, 97
  reflection, 6, 108
  transmission, 6, 188
constant attenuation, 98
cooling system, 258
coupled cells
  stacks of, 188
coupler, 83, 89, 106